Reasoning in Science and Mathematics

*Essays on Logic as
the Art of Reasoning Well*

Richard L. Epstein

Advanced Reasoning Forum

COPYRIGHT © 2012 Richard L. Epstein.

ALL RIGHTS RESERVED. No part of this work covered by the copyright hereon may be reproduced or used in any form or by any means—graphic, electronic, or mechanical, including photocopying, recording, taping, Web distribution, information storage and retrieval systems, or in any other manner—without the written permission of the author.

The moral rights of the author have been asserted.

Names, characters, and incidents relating to any of the characters in this text are used fictitiously, and any resemblance to actual persons, living or dead, is entirely coincidental.

The illustration on page 15 is by Alex Raffi.

For more information visit our website:
    www.AdvancedReasoningForum.org
Or contact us:
    Advanced Reasoning Forum
    P. O. Box 635
    Socorro, NM 87801 USA
    rle@AdvancedReasoningForum.org

ISBN 978-0-9834521-2-6

Reasoning in Science and Mathematics

Essays on Logic as the Art of Reasoning Well

by

Richard L. Epstein

Preface

**Background: Claims, Inferences,
   Arguments, Explanations** 1

Models and Theories 19

Experiments 53

Mathematics as the Art of Abstraction 67

Bibliography 111

Index 119

Essays on Logic as the Art of Reasoning Well

CAUSE and EFFECT, CONDITIONALS, EXPLANATIONS
   Cause and Effect
   The Directedness of Emotions
   Conditionals
   Explanations

PRESCRIPTIVE REASONING
   Reasoning with Prescriptive Claims
   Prescriptive Models
   Rationality

The FUNDAMENTALS of ARGUMENT ANALYSIS (2013)
   Arguments
   Base Claims
   Fallacies
   Subjective Claims
   Generalizing
   Analogies
   Induction and Deduction
   Rationality
   Truth
   Probabilities
   On Metaphysics

REASONING and FORMAL LOGIC (2013)
   Valid Inferences
   The Metaphysical Basis of Logic
   A General Framework for Semantics for Propositional Logics
   Why Are There So Many Logics?
   Truth
   Vagueness
   On Translations
   Three Questions about Logic
   Language, Thought, and Meaning
   Reflections on Some Technical Work in Formal Logic
      Gödel's Theorem
      Categoricity with Minimal Metaphysics
      On the Error in Frege's Proof that Names Denote
   The Twenty-First or "Lost" Sophism on Self-Reference of John Buridan

Preface

This series of books is meant to present the fundamentals of reasoning well in a clear enough manner to be accessible to both scholars and students. The body of each essay gives the main development of the subject, while the footnotes and appendices place the research within a larger scholarly context.

The topic of this volume is the nature and evaluation of reasoning in science and mathematics. Science and mathematics can both be understood as proceeding by a method of abstraction from experience. Mathematics is distinguished from other sciences only in its greater abstraction and its demand for necessity in its inferences.

That methodology of abstraction is the main focus here. In a companion volume, *Cause and Effect, Conditionals, Explanations*, the roles of laws and explanations in science are discussed more fully.

The study of these subjects is not just of academic interest. If we can be clear about the methods and evaluation of science and mathematics, we can clarify our ideas and do better work as scientists and mathematicians. We have a duty to understand what we are doing so that we can not only produce good research but explain that work to others. First comes clear thinking, then comes clear research and clear writing.

* * * * * * * * * *

Background material
In order to make the essays suitable to be read independently, there is some repetition in them, and brief introductions to some background material on claims, inferences, arguments, and explanations are included. Those are filled out in the first essay "Background." That in turn is only a brief sketch of the ideas which are developed in *The Fundamentals of Argument Analysis* and the essay "Explanations" in *Cause and Effect, Conditionals, Explanations*, both in this series of books.

Acknowledgments

Many people have helped me over the many years I have been working on the material in this volume. William S. Robinson and Fred Kroon, in particular, have given much of their time and thought to suggestions that have improved the work. The late Benson Mates was a major inspiration for much of the effort to clarify my ideas.

Charlie Silver, Branden Fitelson, Peter Eggenberger, Stephen Epstein, Carolyn Kernberger, Jack Birner, and the members of the Advanced Reasoning Forum helped me a great deal in the initial stages of the essay "Models and Theories" in 1999–2002; and Greg Miller, David Sherry, and Steffan Angere offered useful comments on this latest version. David Isles, Ian Grant, Paul Livingston, Charlie Silver, Carlo Cellucci, Jeremy Avigad, Reuben Hersh, Andrew Aberdein, Ian Dove, David Sherry, and Steffan Angere helped me improve the essay "Mathematics as the Art of Abstraction." I have benefited, too, from the advice of the editors for the Advanced Reasoning Forum, Michael Rooney and Peter Adams in preparing this volume. LynnDianne Beene did the copyediting which made this a much better book.

Much that is good in this book comes from the generous help of these people, to whom I am most grateful. The mistakes are mine, all mine.

Publishing history of the essays

"Models and Theories" is a revision of "On Models and Theories, with Applications To Economics," *The Bulletin of Advanced Reasoning and Knowledge*, vol. 2, 2004, pp. 79–100, parts of it having appeared previously in *The Guide to Critical Thinking in Economics*, Southwestern, 2004, and *Science Workbook for Critical Thinking*, Wadsworth, 2002.

An earlier version of "Mathematics as the Art of Abstraction" appeared in *Computability: Computable Functions, Logic, and the Foundations of Mathematics*, 3rd ed., Advanced Reasoning Forum, 2008. A later revision appeared in *The Argument of Mathematics*, eds. Andrew Aberdein and Ian Dove, Springer, 2012.

"Experiments" draws on material that first appeared in *Science Workbook for Critical Thinking*, Wadsworth, 2002 and in *Science Reasoning Supplement to the Pocket Guide to Critical Thinking*, Advanced Reasoning Forum, 2011.

The material in the background essay is revised from work in *Critical Thinking*, Wadsworth, 3rd ed., 2005, and *Five Ways of Saying "Therefore"*, Wadsworth, 2001.

Background

This essay presents the essential material about inferences, arguments, and explanations needed for the essays that follow. This is itself only a sketch, drawing on the material in *The Fundamentals of Argument Analysis* and *Cause and Effect, Conditionals, Explanations*.

Claims 1
Inferences 2
Arguments 4
Repairing arguments 7
Generalizations 9
Reasoning backwards 11
Analogies 11
Explanations 12
Arguments and explanations 14
Confirming explanations 16
Inference to the best explanation 17

Claims

Claims A *claim* is a written or uttered part of speech that we agree to view as true or false, but not both.

The word "uttered" is meant to include silent uttering to oneself, what we might call a linguistic thought.

We do not need to make a judgment about whether a sentence is true or whether it is false to classify it as a claim. A claim need not be an *assertion*: a piece of language put forward as true by someone.

Some say that claims only represent things that are true or false: abstract propositions or thoughts. But utterances are what we use in reasoning together, and we can focus on those, as representatives, if you like, of abstract propositions or thoughts.

The word "agree" in the definition of "claim" suggests that it is a matter of convention whether we take a sentence to be a claim. But almost all our conventions, agreements, and assumptions are implicit. Our agreements may be due to many different reasons or causes, including perhaps that there are abstract propositions.

Often when we reason we identify one utterance with another, as when Dick says "Ralph is not a dog" and later, when Suzy thinks about it, she says "I agree. Ralph is not a dog." We do so when we believe that the utterances are equivalent for all our purposes in reasoning.

Equivalent claims To say that wo claims are equivalent for our purposes in reasoning means that no matter how the world could be, the one is true if and only if the other is true.

I will often assume such equivalences without explicitly saying so.

Often what people say is *too vague* to take as a claim: there's no single obvious way to understand the words, as when someone says "This is a free country." Yet, since everything we say is somewhat vague, it isn't whether a sentence is vague but whether it's too vague, given the context, for us to agree to view it as true or false. In an auditorium lit by a single candle, some parts are clearly lit and some are clearly dark, even if we can't draw a precise line where it stops being light and starts being dark. The *drawing the line fallacy* is to argue that if you can't make the difference precise, there's no difference.

Inferences

We reason in order to discern whether certain claims are true. But we also reason to discern whether a particular claim follows from one or more other claims. We might not know whether those other claims are true. But were they true, would the truth of this other claim follow?

Inferences An *inference* is a collection of claims, one of which is designated the *conclusion* and the others the *premises*, which is intended by the person who sets it out either to show that the conclusion follows from the premises or to investigate whether that is the case.

In order to investigate the idea of a conclusion following from the premises of an inference, we make some definitions.

Valid, strong, and weak inferences An inference is *valid* if it is impossible for the premises to be true and conclusion false at the same time and in the same way.

An inference is ***strong*** if it is possible but unlikely for the premises to be true and conclusion false at the same time and in the same way. An inference is ***weak*** if it is neither valid nor strong.

The classification of invalid inferences is on a scale from the strongest to the weakest, as we deem the possibilities we consider in which the premises are true and conclusion false to be more or less likely.

The following, for example, is a valid inference:

> Maria is a widow.
> So Maria was married.

We do not know if the premise is true, but if it is, then the conclusion is not false. In this case the conclusion surely follows from the premise.

The following is valid, too:

> All dogs bark.
> Spot is a dog.
> So Spot barks.

Here we know that the first premise is false: Basenjis can't bark, and some dogs have had their vocal cords cut. It's not the truth or falsity of the premises and conclusion that determines whether an inference is valid; rather, it is the ways in which the premises and conclusion could be true or false. Were the premises of this inference true, the conclusion would be also; the inference is valid.

In contrast, the following inference is strong:

> Almost all dogs bark.
> Ralph is a dog.
> So Ralph barks.

If we know no more about Ralph than that he is a dog, then any way in which the premises could be true and conclusion false is unlikely, for we know how rare those are. In this case, too, we say that the conclusion follows from the premises, though there is no certainty, no "must" in that. It is only that, relative to what we know, it seems to us very unlikely that the premises could be true and conclusion false.

The following, however, is weak:

> Louise is a student.
> So Louise isn't married.

There lots of ways the premise could be true and conclusion false: for all we know, Louise might be forty years old with a husband and child.

Our evaluation of the strength of an inference is relative to what we believe. "Likely" means "likely to us." But typically the scale from strong to weak is not so completely relative to a particular person that there is no hope we can agree on the strength of inferences. Suppose we disagree. I find a particular inference strong, and you find it weak. If we wish to reason together, you should describe to me a way the premises could be true and the conclusion false that you think is not unlikely. That may depend on knowledge you have of how the premises could be true that I do not have, but once you've made that explicit, we can agree or disagree that there is such a possibility. The only issue, then, would be whether we agree that the possibility is likely. Sometimes we can't come to a clear determination, but further examination will leave us with a clearer understanding of what our differences in evaluation are, based on more than just whim. When the beliefs involved in determining the strength of an inference are made explicit, determining the inference to be strong or weak is far more likely to be a shared judgment.

In sum, we say that the conclusion of an inference *follows from* the premises if the inference is valid or strong.

Arguments

The paradigmatic use of inferences is attempts to convince someone that a claim is true.

Arguments An *argument* is an inference that is intended by the person who sets it out to convince someone that the conclusion is true.

Arguments are attempts to convince, whether someone tries to convince you, or you try to convince someone else, or you try to convince yourself. But that does not mean that the criterion for whether an argument is good is whether the argument actually does convince. If your friend is drunk, you may give him an excellent argument that his driving home is dangerous; though he remains unconvinced, the argument is no worse. A politician may make a bad argument that you should vote for her, but though you may be convinced that does not mean the argument is good. Perhaps other ways to convince, such as entreaties, exhortations, sermons, or advertisements can be judged by how well they convince, but that is not a criterion for judging attempts to establish the truth of a claim. A *good argument* is one that gives us good reason to believe the conclusion. But what does "good reason" mean?

If an argument is to give us good reason to believe its conclusion, we should have good reason to believe its premises, for from a false claim we can reason as easily to a false conclusion as a true one.

> The Prime Minister of England is a dog. All dogs have fur.
> So the Prime Minister of England has fur. (false conclusion)
>
> The Prime Minister of England is a dog. All dogs have a liver.
> So the Prime Minister of England has a liver. (true conclusion)

It seems, then, that a good argument should have true premises. But consider:

> There are an even number of stars in the sky.
> So the number of stars in the sky can be divided by 2.
>
> There are an odd number of stars in the sky.
> So the number of stars in the sky cannot be divided by 2.

One of these has a true premise, but we cannot tell which. A standard that gives us no way to evaluate arguments is not part of the art of reasoning well. Rather, for an argument to be good we must have good reason to believe its premises and recognize that we have good reason to believe them, and as well actually believe them, for otherwise what convincing is done has no basis in our beliefs.

Plausible claims A claim is *plausible* to a particular person at a particular time if:
- The person has good reason to believe it.
- The person recognizes that he or she has good reason to believe it.
- The person believes it.

A claim that is not plausible is *implausible or dubious*.

The classification of claims as plausible or implausible is on a scale from the most plausible, ones we recognize as true, to the least plausible, ones we recognize as false. Though we do not have precise measures of plausibility, we can often compare the plausibility of claims; and by being explicit about our background we can usually agree on whether we will take any particular claim to be plausible. If we did not think that we can share our judgments of what is plausible, we would have no motive for trying to reason together. So if I say a claim is *plausible* without specifying a particular person, I mean it's plausible to most of us now as I'm writing.

Good reason to believe a claim cannot always be established by reasoning. We must have some place to start, some plausible claims that require no further justification, or we could never get started. Some claims we take as plausible because of our personal experience, or our trust in authority, or our beliefs about the nature of the world.

But it's not just that the premises of a good argument have to be plausible. They have to be more plausible than the conclusion, for otherwise they would give us no more reason to believe the conclusion than we had without the argument.

Begging the question An argument *begs the question* if it has a premise that is not more plausible than its conclusion.

Further, for an argument to be good, the conclusion must follow from the premises. For example, consider:

Richard L. Epstein is the author of this essay.
So Richard L. Epstein is bald.

This argument is weak: there are lots of likely ways the premise could be true and conclusion false. Though you know the premise is true, that gives no reason to believe the conclusion. We classify arguments, since they are inferences, as valid, strong, or weak, and unless an argument is valid or strong the conclusion doesn't follow from the premises. But do strong arguments give good reason to believe the conclusion?

Consider that last week Dick heard there are parakeets for sale at the mall. He knows that his neighbor has a birdcage in her garage, and he wonders whether the cage will be big enough for one of them. He reasons:

(‡) Every parakeet I or anyone I know has seen, or read, or heard about is less than 50 cm tall. So the parakeets on sale at the mall are less than 50 cm tall.

This argument isn't valid. A new kind of parakeet that is 1 meter tall might have been discovered in the Amazon; or a new supergrow bird food has been developed that makes parakeets grow very tall; or aliens have captured some parakeets and hit them with rays to make them very large; or . . . . But any possibility Dick or we can think of for the premise to be true and conclusion false is unlikely—so unlikely that Dick and we have good reason to believe the conclusion. The argument is strong. It gives us good enough reason to believe the conclusion. Moreover, it is better than a valid one with the same conclusion. Replacing the premise of (‡) with "All parakeets are less than 50 cm tall" would yield a valid argument, but a worse one, for that claim is less plausible than the premise of (‡). There is often a trade-off between how plausible the premises of an argument are and how strong the argument is: the less plausible the premises, the stronger the argument.

A strong argument can give us good reason to believe its conclusion, at least in our daily lives, and in science, too, as we shall see.

We now have three tests an argument to pass for it to be good.

Necessary conditions for an argument to be good
- The premises are plausible.
- The premises are more plausible than the conclusion.
- The argument is valid or strong.

These conditions are relative to a particular person, though in practice we can have confidence that they establish an intersubjective standard for the evaluation of arguments.

Whether these conditions are also sufficient is a large topic that I examine in *The Fundamentals of Argument Analysis*. In what follows, I will generally treat them as both necessary and sufficient.

Repairing arguments

In our daily lives we often treat arguments as good that don't satisfy these conditions. For example, consider:

    Lee: Tom wants to get a dog.
    Maria: What kind?
    Lee: A dachshund. And that's really stupid, since he wants one that will chase a frisbee.

Lee has made an argument if we interpret what he said: Tom wants a dog that will chase a frisbee, so Tom shouldn't get a dachshund. On the face of it that argument is not strong or valid. Still, Maria knows very well, as do we, that a dachshund would be a bad choice for someone who wants a dog to play with a frisbee: dachshunds are too low to the ground, they can't run fast, they can't jump, they'd trip over a frisbee so they couldn't bring it back. Any dog like that is a bad choice for a frisbee partner. Lee left out these obvious claims. But why should he bother to say them?

We normally leave out so much that if we look only at what is said, we will be missing too much. We can and should rewrite many arguments by adding an unstated premise or even an unstated conclusion.

When are we justified in doing so? How do we know whether we've rewritten an argument well or just added our own assumptions? To repair arguments that are apparently defective, we must have some standards, for otherwise we will end up putting words in other folks' mouths. Such standards depend on what we can assume about the person with whom we are reasoning or whose work we are reading.

The Principle of Rational Discussion   We assume that the other person with whom we are deliberating or whose reasoning we are evaluating:
- Knows about the subject under discussion.
- Is able and willing to reason well.
- Is not lying.

Often someone with whom we wish to reason does not satisfy these conditions. But when we discover that, then it makes no sense to continue to reason with him or her. We should be educating, or consoling, or pointing out errors. The Principle of Rational Discussion justifies adopting the following guide.

The Guide to Repairing Arguments   Given an (implicit) argument that is apparently defective, we are justified in *adding* one or more premises or a conclusion if and only if all the following hold:
- The argument becomes valid or strong.
- The premise is plausible and would seem plausible to the other person.
- The premise is more plausible than the conclusion.

If the argument is valid or strong, yet one of the original premises is implausible, we may *delete* that premise if the argument becomes no worse. In that case we say the premise is ***irrelevant***.

Given only this Guide, we might try to repair every argument into a good one. That would be wrong, for there are standards for when an argument is unrepairable.

Unrepairable Arguments We cannot repair a (purported) argument if any of the following hold:
- There is no argument there.
- The argument is so lacking in coherence that there's nothing obvious to add.
- A premise is implausible or several premises together are contradictory and cannot be deleted.
- The obvious premise to add would make the argument weak.
- Any obvious premise to add to make the argument strong or valid is implausible.
- The conclusion is clearly false.

It's not that when we encounter one of these conditions we can be sure the speaker had no good argument in mind. Rather, we are not justified in making that argument for him or her, for it would be putting words in their mouth.

In addition to these conditions for an argument to be unrepairable, a list of other kinds of arguments, called *fallacies*, have been deemed to be typically so bad that they, too, are rejected as unrepairable.

Consideration of two particular kinds of arguments is important for the essays that follow.

Generalizations

Generalizations A *generalization* is an argument in which we conclude a claim about a group, called the ***population***, from a claim about some part of it, the ***sample***. Sometimes we call the conclusion the *generalization*. Plausible premises about the sample are called the ***inductive evidence*** for the generalization.

The following are generalizations:

(‡‡) Every dog I've seen barks.
So all dogs bark.
sample: The dogs the speaker has seen.
population: All dogs.

Every dog I ever met except one can bark.
So almost all dogs bark.
sample: The dogs the speaker has met.
population: All dogs.

Of dog owners who were surveyed, 98.2% said their dogs bark.
So about 98% of all dogs bark.
sample: The dogs of the pet owners surveyed.
population: All dogs.

The last is called a *statistical generalization* because its conclusion is a statistical claim about the population.

A generalization is worthless, that is, a bad argument, if we have no reason to think that the sample is similar to the population. What we want for a good generalization is for the sample to be representative.

Representative sample A sample is *representative* if no one subgroup of the whole population is represented more than its proportion in the population. A sample is **biased** if it is not representative.

The first and second examples at (‡‡) are bad because we have no reason to think that the dogs the speaker has seen are representative of all dogs. We don't know enough about the sample in the third generalization to make a judgment about whether it's representative.

Random sampling is an important method for getting a representative sample.

Random sampling A sample is *chosen randomly* if at every choice there is an equal chance for any of the remaining members of the population to be picked.

Random sampling does not guarantee that the sample will be representative. Choosing two students randomly from the 716 at McEpstein High School to interview about their views on gay marriage is not going to give a representative sample. The sample has to be large enough for us to have good reason to think it is representative.

But even if we have confidence that the sample is representative, if it's not studied well then it's no use for concluding anything about the population. Maria personally asked all but three of the thirty-six people in her criminology class whether they've ever used cocaine, and only two said yes. So she concluded that almost no one in the class has used cocaine. But there's no reason to think that people will answer truthfully to such a question, so her generalization is not good.

Necessary conditions for a generalization to be good
- The sample is representative.
- The sample is big enough.
- The sample is studied well.

Reasoning backwards

One particular mistake in reasoning is important to note for some of the discussions that follow. As an example, Suzy said to Lee:

All CEOs of computer companies are rich. Bill Gates is a CEO of a computer company. So Bill Gates is rich.

Lee sees that Suzy's argument is valid, and he knows that Bill Gates is rich and that he's a CEO of a computer company. So he reckons that "All CEOs of computer companies are rich" is true, too. But he's wrong: there are lots of CEOs of small, struggling computer companies who are not rich. Lee is arguing backwards.

Arguing backwards *Arguing backwards* is the mistake of concluding that the premises of an inference are true because the inference is valid or strong and its conclusion is plausible.

Analogies

One particular kind of argument proceeds by making a comparison.

Analogies A comparison becomes *reasoning by analogy* when it is part of an argument: On one side of the comparison we draw a conclusion, so on the other side we say we can conclude the same.

For example, consider:

> We should legalize marijuana. After all, marijuana is just like alcohol and tobacco, and those are legal.

The comparison here is between marijuana on the one hand and alcohol and tobacco on the other. The latter are legal. So we should make marijuana legal, too. But no connection between the premise and the conclusion has been supplied other than saying that marijuana "is just like" alcohol and tobacco. Making that clear, saying in what ways marijuana is like alcohol and tobacco, and then stating a general principle that invokes those similarities is what is needed. In doing so, whatever differences there are between marijuana, alcohol, and tobacco are implicitly if not explicitly claimed not to matter.

The difficulty in reasoning by analogy is to make clear what we mean by "is just like" or "is the same as" in order to justify the inference in terms of the comparison. Such a justification calls for some general claim under which the two sides of the comparison fall. Often analogies are sketchy, with only the comparison offered, so that their main value is to stimulate us to search for such a general claim. It must be one that relies on the similarities and for which the differences between the two sides of the comparison don't matter. Though that procedure is somewhat more involved than in analyzing many other arguments, it is does not involve further conditions for an argument to be good.

Besides arguments, there is another use of inferences that is important in science.

Explanations

Explanations that are meant to answer why some claim is true can be understood as inferences.

> Why is the sky blue? Because sunlight is refracted through the atmosphere so as to absorb other wavelengths of light.

"The sky is blue" is explained in terms of why it's true, what it follows from, the reasons for its truth.

Inferential explanations An *explanation* is a collection of claims that can be understood as E because of A, B, C, . . . that is meant to answer the question "Why is E true?"

What are the conditions for an inferential explanation to be good? We can't explain what's not plausible. For example, we can't explain why "All dogs hate cats" is true because we know that's false.

Since what's being explained is meant to follow from the claims doing the explaining, the explanation should be valid or strong. The explanation "Dogs lick their owners because dogs aren't cats" is not good because that inference is neither valid nor strong and there's no obvious way to repair it.

As well, the claims doing the explaining should be plausible. We don't accept "The sky is blue because there are blue globules high in the atmosphere" as a good explanation because we know that's false.

But how plausible do they need to be? Consider:

Dick: Ohhh. My head hurts.
Zoe   You drank three cocktails before dinner, a bottle of wine with dinner, then a couple of glasses of brandy. Anyone who drinks that much is going to get a headache.

Zoe has given Dick an explanation of why he has a headache:

Anyone who drinks that much is going to have a headache.
Therefore (explains why), Dick has a headache.

It's a good explanation. But judged as an argument, it's bad, for it begs the question: it's a lot more obvious to Dick that he has a headache than that anyone who drinks that much is going to have a headache. In a good explanation at least one of the premises is not more plausible than the conclusion, for otherwise it would be an argument; but we already have reason to believe the conclusion, so we need no argument for that. An explanation is meant to help us understand why, to place our knowledge in a clearer context. It is not meant to convince.

Nor can we explain why a claim is true by just restating the claim in other words. "Dick is having trouble writing his essay today because he has writer's block" is a bad explanation because having writer's block just means not being able to write.

Necessary conditions for an inferential explanation to be good
For an inferential explanation "E because of A, B, C, . . ." to be good, all the following must hold:

- E is highly plausible.

- Each of A, B, C, ... is plausible, but at least one of them is not more plausible than E.
- "A, B, C, ... therefore E " is a valid or strong inference, possibly with respect to some plausible unstated claims.
- The explanation is not "E because of D" where D is E itself or a simple rewriting of E.

These are necessary though not obviously sufficient conditions for an explanation to be good. Other conditions have been proposed as necessary, too, but there is little agreement on them.

Arguments and explanations

Dick, Zoe, and Spot are out for a walk in the countryside. Spot runs off and returns after five minutes. Dick notices that Spot has blood around his muzzle. And Zoe and he both really notice that Spot stinks like a skunk. Dick turns to Zoe and says, "Spot must have killed a skunk. Look at the blood on his muzzle. And he smells like a skunk."

Dick has made a good argument:

Spot has blood on his muzzle. Spot smells like a skunk.
Therefore, Spot killed a skunk.

Dick has left out some premises he knows are as obvious to Zoe as to him:

Spot isn't bleeding.
Skunks aren't able to fight back very well.
Dogs try to kill animals by biting them.
Normally when Spot draws a lot of blood from an animal that's smaller than him, he kills it.
Only skunks give off a characteristic skunk-odor that drenches whoever or whatever is near if they are attacked.

Zoe replies, "Oh, that explains why he's got blood on his muzzle and smells so bad." That is, she takes the same claims and views them as an explanation, a good explanation, relative to the same unstated premises:

Spot killed a skunk
explains why Spot has blood on his muzzle and smells like a skunk.

For an explanation "E because of A, B, C, ...", we can ask what evidence we have for A. Sometimes we can supply all the evidence we need just by reversing the inference. For Zoe's explanation to be good, "Spot killed a skunk" must be plausible, and it is because of the argument Dick gave—they needn't wait until they find the dead skunk.

Explanations and associated arguments For an inferential explanation:

A, B, C, ...
Therefore$_{\text{explanation}}$ E

the ***associated argument*** to establish A is:

E, B, C, ...
Therefore$_{\text{argument}}$ A

An explanation is ***dependent*** if one of the premises is not plausible and the associated argument for that premise is not good. An explanation is ***independent*** if it is not dependent.

If an explanation is dependent, then it lacks evidence for at least one of its premises which can't be supplied by an associated argument. Each premise of an independent explanation is plausible, either because of the associated argument for it or for independent reasons, such as our knowing that most dogs bark or that Sheila is not a herring. Still, an independent explanation might be bad. After all, "Dogs lick their owners because dogs aren't cats" has a premise that's clearly true.

Confirming explanations

> Flo: Spot barks. And Wanda's dog Ralph barks. And Dr. E's dogs Anubis and Juney bark. So all dogs bark.
>
> Barb: Yeah. Let's go over to Maple Street and see if all the dogs there bark, too.

Flo, who's five, is generalizing. Her friend Barb wants to test the generalization.

Suppose that A, B, C, D are given as inductive evidence for a generalization G. Some other plausible unstated premises may also be needed, but let's keep those in the background. Then we have that G explains A, B, C, D.

But if G is true, we can see that it follows that some other claims are true, instances of the generalization G, say L, M, N. If those are true, then G would explain them, too (Fido barks, Lady barks, Buddy barks). That is, G explains A, B, C, D and predicts L, M, N, *where the difference between the explanation and the prediction is that we don't know if the prediction is true*, not that the prediction is about the future.

Suppose we find that L, M, N are indeed true. Then the argument A, B, C, D + L, M, N, therefore G is a better one for G than we had before. At the very least it has more instances of the generalization as premises.

How can more instances of a generalization prove the generalization better? They can if (1) they are from different kinds of situations, that is, A, B, C, D + L, M, N cover a more representative sample of possible instances of G than do just A, B, C, D. Typically that's what happens: we deduce claims from G for situations we had not previously considered. And (2) because we had not previously considered the kind of instances L, M, N of the generalization G, we have some confidence that we haven't got G by manipulating the data, selecting situations that would establish just this hypothesis.

A good way to test an hypothesis or generalization is to try to falsify it. Trying to falsify a generalization means we are trying to come up with instances of the generalization to test that are as different as we can imagine from the ones we first used in deducing it. Trying to falsify is just a good way to ensure (1) and (2). We say that an experiment ***confirms***—to some extent—a particular doubtful claim in the premises if it shows that a prediction is true.

The story is much the same for claims that aren't generalizations. If A, B, C, D, E provide an argument for G, and G explains (has as consequences) L, M, N, O, P, we can check whether those are true. If they are, it is often the case that A, B, C, D, E + L, M, N, O, P provide a stronger proof of G. On the other hand, if one of those turns out to be false, then G is (most likely) false. *Confirming an hypothesis-explanation is just strengthening the associated argument.*

Inference to the best explanation

> It can hardly be supposed that a false theory would explain, in so satisfactory a manner as does the theory of natural selection, the several large classes of facts above specified [the geographical distribution of species, the existence of vestigial organs in animals, etc.]. It has recently been objected that this is an unsafe method of arguing; but it is a method used in judging of the common events of life, and has often been used by the greatest natural philosophers.
> Charles Darwin, *On the Origin of Species*, p. 476

If Darwin was right, why did scientists spend the next hundred years trying to confirm or disprove the hypotheses of natural selection? Darwin is arguing backwards: from the truth of the conclusion(s), he infers the truth of the premises. Rather, the evidence for the claims doing the explaining comes from strengthening the associated argument.

We don't have accepted criteria for what counts as the best explanation, and in any case it's only the best explanation we've thought of so far. Scientists have high hopes for their hypotheses and are motivated to investigate them if they appear to provide a better explanation than current theories. But the scientific community quickly corrects anyone who thinks that just making an hypothesis that would explain a lot establishes that it's true.

Fallacy of inference to the best explanation The *fallacy of inference to the best explanation* is to argue that because some claims constitute the best explanation we have, they're therefore true.

This concludes the very brief summary of the basics of analysis of inferences, arguments, and explanations assumed as background in the succeeding essays.

Models and Theories

Models and theories arise by a process of abstraction. They do not codify truths of the world but rather set out ways in which to reason about the world through ignoring what we consider to be extraneous. By better understanding the process of abstraction, we can proceed more clearly in creating and evaluating theories. Two case studies illustrate the evaluation of theories in terms of abstraction.

Analogy and abstraction 19
Examples of models and theories
    1. A map of Minersville 20
    2. Models of the solar system 21
    3. The kinetic theory of gases 24
    4. The acceleration of falling objects 26
    5. Newton's laws of motion and
       Einstein's theory of relativity 27
    6. Ether as the medium of propagation of light waves . . . 27
    7. Euclidean geometry 27
    8. Classical propositional logic 28
    9. Electrical switches 29
Models, theories, and truth 30
Theories and confirmation 31
Modifying theories in the light of the evidence 33
Case study: A model with no path of abstraction 38
Case study: Confusing truth with applicability 43
Notes 46

Analogy and abstraction

A comparison becomes ***reasoning by analogy*** when it is part of an argument: On one side of the comparison we draw a conclusion, so on the other side we say we should conclude the same.

    The difficulty in reasoning by analogy is to make clear what we mean by "is just like" or "is the same as" in order to justify the inference in terms of the comparison. Such a justification calls for some general claim under which the two sides of the comparison fall. Often analogies are sketchy, with only the comparison offered, so that their main value is to stimulate us to search for such a general claim. That claim must be one that relies on the similarities and for which the differences between the two sides of the comparison don't matter.

We cannot nor do we want to pay attention to all we encounter. We take some parts of our experience to be more important than others, and we reason from those as if they were all that mattered. When we do this consciously, when we say that we'll pay attention to only this part and ignore all the rest, we are *abstracting from experience*. When we say that what we have ignored doesn't matter in drawing conclusions from our experience, we are *reasoning by abstraction*. We can understand models and theories in this way: we ignore much and then can draw conclusions when the differences don't matter.

Examples of models and theories

Example 1   A map of Minersville, Utah—reasoning by analogy
On the facing page is an accurate map of Minersville, Utah. Looking at it we can see that the streets are evenly spaced. For example, there is the same distance between 100 N and 200 N as between 100 E and 200 E. The last street to the east is 300 E. There is no paved road going north beyond Main Street on 200 E.

That is, from this map we can deduce claims about Minersville, even if we've never been there. But there is much we can't deduce: Are there hills in Minersville? Are there lots of trees? How wide are the streets? How far apart are the streets? Where are there houses? The map is accurate for what it pays attention to: the relative location and orientation of streets. But it tells us nothing about what it ignores.

The differences between the map and Minersville aren't important when we infer that the north end of 200 W is at 200 N. In contrast, a scale model of a city or a mountain abstracts less from the actual terrain: height and perhaps placement of rivers and trees are there. The map of Minersville *abstracts more* from the actual terrain than a scale model of the city would, that is, *it ignores more*.

To use this model is to reason by analogy: we can draw conclusions when appropriate similarities are invoked and the differences do not matter. The general principle, in this example, is not stated explicitly. The discussion above suggests how we might formulate one, but it hardly seems worth the effort. We can "see" when someone has used a map well or badly.[1]

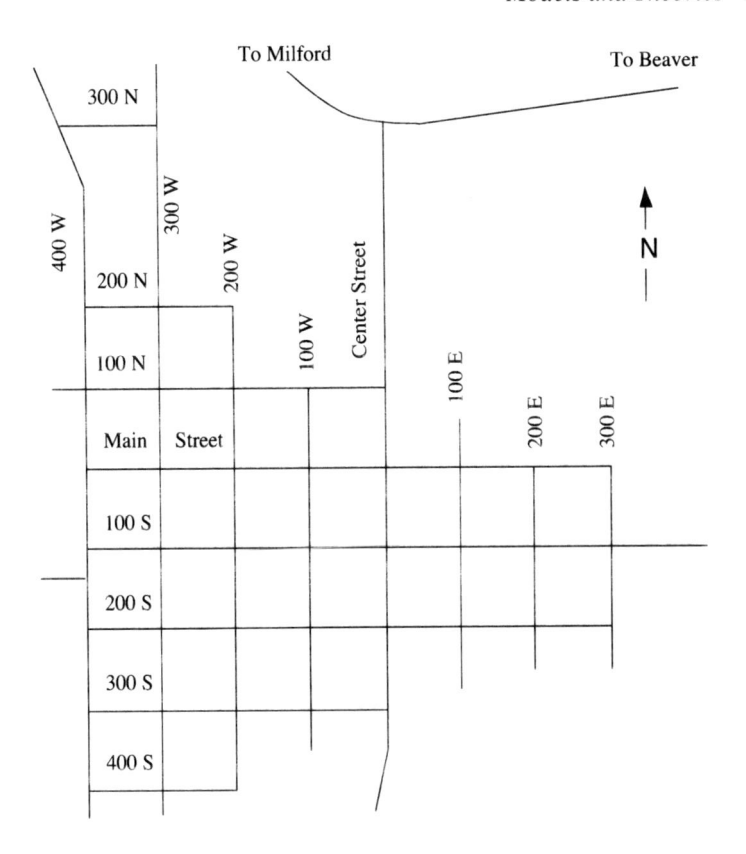

Example 2 Models of the solar system—choosing between models
On the next page is a sketch of the model of the universe that Ptolemy proposed in Egypt in the second century A.D. It's meant to show the relative positions of the planets, sun, and moon, and the ways they move. We can't deduce anything about, say, the size of the planets, the distances between them, nor the speeds at which they move because this model ignores those. According to this model, each of the moon, sun, and planets revolves around the Earth in a circular orbit, all moving in the same direction. Along that orbit, each planet also revolves in a smaller circle, called an "epicycle." The sun, Earth, and Venus are always supposed to be in a line as shown in the picture.

Ptolemy made a lot more claims about the planets, Earth, and sun that were to be used in making predictions, but for our purposes this sketch will do.

22   *Reasoning in Science and Mathematics*

Ptolemy's model

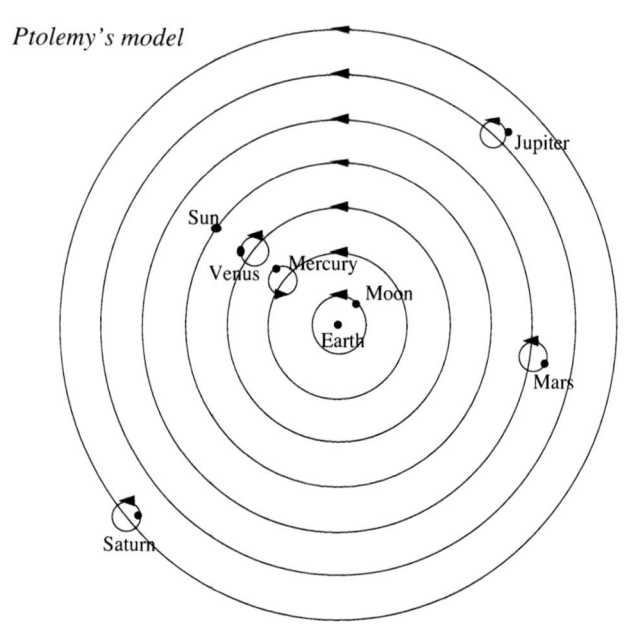

Ptolemy's model accorded pretty well with observations of the movements of the planets and was the generally accepted way to understand the universe for many centuries. But in 1543 the Polish astronomer Copernicus published a book with a different model of the universe.

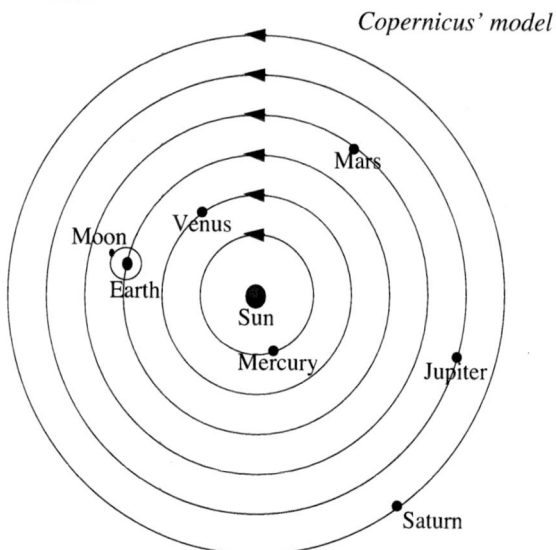

Copernicus' model

This sketch, too, abstracts a lot from what is being modeled. The sun is shown to be larger than the planets, but that's all we see about their relative sizes. We can't tell from the picture whether the orbits are all on the same plane or on different planes. We do see that the planets all revolve in the same direction, and that the Earth, sun, and Venus do not always stay lined up.

Ptolemy accounted for the motion of the sun, planets, and stars in the sky by saying they revolved around the Earth every 24 hours. Copernicus accounted for those motions by saying that the Earth rotated around its own axis every 24 hours. How could someone in the late 16th century decide between these two models? Both were in accord with the observations that had been made.

In the early 1600s the telescope was invented, and in 1610 Galileo built his own telescope with a magnification of about 33 times, using it to study the skies. One of his students suggested an experiment that might distinguish between the Ptolemaic and Copernican models. Venus was too far from the Earth to be seen as anything other than a spot of light. But according to Ptolemy's model, viewed from the Earth at most only a small crescent-shaped part of Venus will be illuminated by the sun. From Copernicus' model, however, we can deduce that from the Earth Venus should go through all the phases of illumination, just like the moon: full, half, crescent, dark, and back again. Galileo looked at Venus through his telescope for a period of time and saw that it exhibited all phases of illumination, and this he took to be proof that Copernicus' model was correct.

Not a lot of other people were convinced, however. Telescopes were rare and not very reliable: they introduced optical illusions, such as halos, from the imperfections in the glass and the mounting. Why should astronomers have trusted Galileo's observations?

It was more due to Newton that something like Copernicus' model of the universe was finally accepted. Newton deduced from his laws of motion that the orbits of the Earth, sun, and the planets would have to be ellipses, not circles. And the distances between them would have to be much greater than supposed. Using Newton's laws, Edmond Halley predicted correctly the return of a comet that had been observed in 1682. Telescopes were better, with fewer optical illusions, and they were common enough that most astronomers could use one, so better and better observations of the planets and stars could be made. Those observations could be deduced from the Copernican-Newtonian model, while new epicycles had to be invented to account for them in the Ptolemaic model.

Note that each model is supposed to be similar to the universe in only a few respects, ones that would have an effect on how we could see the objects in the universe from the Earth. Differences, such as whether Venus is rocky or gaseous, are not supposed to matter for those observations. If the model is correct, then reasoning by analogy—very precise analogy—certain claims can be deduced.

Example 3 The kinetic theory of gases—getting true predictions doesn't mean the model is true

This theory is based on the following postulates, or assumptions.

1. Gases are composed of a large number of particles that behave like hard, spherical objects in a state of constant, random motion.

2. The particles move in a straight line until they collide with another particle or the walls of the container.

3. The particles are much smaller than the distance between the particles. Most of the volume of a gas is therefore empty space.

4. There is no force of attraction between gas particles or between the particles and the walls of the container.

5. Collisions between gas particles or collisions with the walls of the container are perfectly elastic. Energy can be transferred from one particle to another during a collision, but the total kinetic energy of the particles after the collision is the same as it was before the collision.

6. The average kinetic energy of a collection of gas particles depends on the temperature of the gas and nothing else.

J. Spencer, G. Bodner, and L. Rickard, *Chemistry*

Here is a picture of what is supposed to be going on in a gas in a closed container. The molecules of gas are represented as dots, as if they were hard spherical balls.

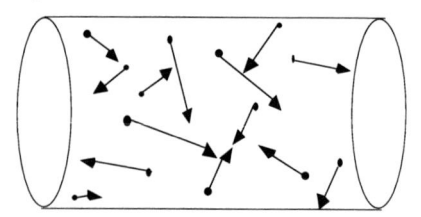

The length of the line emanating from a particle models the particle's speed; the arrow models the direction the particle is moving. The kinetic energy of a particle is defined in terms of its mass and velocity: kinetic energy = .5 mass x velocity^2. The model defines what is meant for a collision to be elastic. In contrast, here is a picture of what happens in an inelastic collision between a rubber ball and the floor.

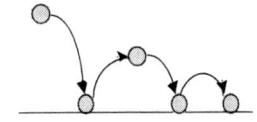

Each time the ball hits the ground, some of its kinetic energy is lost either through being transferred to the floor or in compressing the ball.

What are we to make of these assumptions? Some are false: molecules of gas are not generally spherical and are certainly not solid; the collisions between molecules and the walls of a container or each other are not perfectly elastic; there is some gravitational attraction between the particles and each other and also with the container. How can we use false claims in a model?

The model proceeds by abstraction, much like in analogy: To the extent that we can ignore how molecules of gases are not spherical, and ignore physical attraction between molecules, and ignore . . . we can draw conclusions that may be applicable to actual gases. To the extent that the differences between actual gases and the abstractions don't matter, we can draw conclusions. But how can we tell if the differences matter?

The model suggests that the pressure of a gas results from the collisions between the gas particles and the walls of the container. So if the container is made smaller for the same amount of gas, the pressure should increase; and if the container is made larger, the pressure should be less. So the pressure should be proportional to the inverse of the volume of the gas. That is, the model suggests a claim about the relationship of pressure to volume in a gas. Experiments can be performed, varying the pressure or volume, and they are close to being in accord with that claim.

Other laws are suggested by the model: Pressure is proportional to the temperature of the gas, where the temperature is taken to be the average kinetic energy of the gas. The volume of the gas should be proportional to the temperature. The amount of gas should be proportional to the pressure. All of these are confirmed by experiment.

26   Reasoning in Science and Mathematics

Those experiments confirming predictions from the model do not mean the model is more accurate than we thought. Collisions still aren't really elastic; molecules aren't really hard spherical balls. The kinetic theory of gases is a model that is useful, as with any analogy, when the differences don't matter.

Example 4   The acceleration of falling objects—an equation can be a model

Galileo argued that falling objects accelerate as they fall: they begin falling slowly and fall faster and faster the farther they fall. He didn't need any mathematics to show that. He just noted that a heavy stone dropped from 6 feet will drive a stake into the ground much farther than if it were dropped from 6 inches.

He also said that the reason a feather falls more slowly than an iron ball when dropped is because of the resistance of air. He argued that at a given location on the Earth all objects should fall with the same acceleration. He claimed that the distance traveled by a falling object is proportional to the square of the time it travels. Today, from many measurements, the equation is given by:

(*) $d = \frac{1}{2} (9.82 \text{ meters/sec}^2) \cdot t^2$ where t is time in seconds

The equation (*) is a model by abstraction: we ignore air resistance and the shape of the object, considering only the object's mass and center of gravity. If the differences don't matter, then a calculation from the equation, which is really a deduction, will hold. But often the differences do matter. Air resistance can slow down an object: if you drop a cat from an airplane, it will spread out its legs and reach a maximum velocity when the force of the air resistance equals the force of acceleration.

With this model there is no visual representation of that part of experience that is being described. There is no point-to-point conceptual comparison, nor are we modeling a static situation. The model is couched in the language of mathematics; equations can be models, too.[2]

Example 5   Newton's laws of motion and Einstein's theory of relativity—how a false theory can be used

Newton's laws of motion are taught in every elementary physics course and are used daily by physicists. Yet modern physics has replaced

Newton's theories with Einstein's and quantum mechanics. Newton's laws, physicists tell us, are false.

But can't we say that Newton's laws are correct relative to the quality of measurements involved, even though Newton's laws can't be derived from quantum mechanics? Or perhaps they can if a premise is added that we ignore certain small effects. Yet how is that part of a theory?

A theory is a schematic representation of some part of the world. We draw conclusions from the representation (we calculate or deduce). The conclusion is said to apply to the world. The reasoning is legitimate so long as the differences between the representation and what is being represented don't matter. Newton's laws of motion are "just like" how moderately large objects interact at moderately low speeds; we can use those laws to make calculations so long as the differences don't matter. Some of the assumptions of that theory are used as conditions to tell us when the theory is meant to be applied.[3]

*Example 6 Ether as the medium of light waves — a prediction
can show that an assumption of a theory is false*
In the 19th century light was understood as waves. In analogy with waves in water or sound waves in the air, a medium was postulated for the propagation of light waves: the ether. Using that assumption, many predictions were made about the path and speed of light in terms of its wave behavior. Attempts were then made to isolate or verify the existence of an ether. The experiments of Michaelson and Morley showed those predictions were false. When a better theory was postulated by Einstein, one which assumed no ether and gave as good or better predictions in all cases where the ether assumption did, the theory of ether was abandoned.

Example 7 Euclidean geometry — a model that can't be true
Euclidean plane geometry speaks of points and lines: a point is location without dimension, a line is extension without breadth. No such objects exist in our experience. But Euclidean geometry is remarkably useful in measuring and calculating distances and positions in our daily lives.

Points are abstractions of very small dots made by a pencil or other implement. Lines are abstractions of physical lines, either drawn or sighted. So long as the differences don't matter, that is, so long as the size of the points and the lines are very small relative to what is being measured or plotted, we can rely on whatever conclusions we draw.

No one asks (anymore) whether the axioms of Euclidean geometry are true. Rather, when the differences don't matter, we can calculate and predict using Euclidean geometry. When the differences do matter, as in calculating paths of airplanes circling the globe, Euclidean plane geometry does not apply, and another model, geometry for spherical surfaces, is invoked.

Euclidean geometry is a deductive theory: a conclusion drawn from the axioms is accepted only if the inference is valid. It is a purely mathematical theory, which, taken as mathematics, would appear to have no application since the objects of which it speaks do not exist. But taken as a model it has applications in the usual sense, reasoning to conclusions when the differences don't matter.

Some people, however, invoke points and lines as actual things, abstract, non-sensible things, which then provide the other side of the analogy. To reason solely by abstraction without positing the other side for an analogy seems to impede our ability to imagine clearly. There is no harm in such imaginings, so long as we are clear that the abstract things of which we speak are nothing more than what is left when we ignore much of our experience; they are meant to fill out our picture.

Example 8   Classical propositional logic—a prescriptive model
Suppose we ignore everything about claims except whether they are true or false and how they are built up as compounds using "and," "or," and "not." We can use the symbols &, v, and ~ to stand for these abstracted versions of the connectives. Then the following tables summarize the usual classical understanding of how the truth-value of a compound relates to the truth-value of its parts, where A and B stand for any claims, T stands for "true," and F stands for "false."

A	B	A & B		A	B	A v B		A	~A
T	T	T		T	T	T		T	F
T	F	F		T	F	F		F	T
F	T	F		F	T	F			
F	F	F		F	F	F			

There is also a table for "if . . . then . . .", but these tables are enough for our discussion. Using them repeatedly we can calculate the truth-value of any compound claim built using "and," "or," and "not"

Models and Theories

if we know the truth-values of its parts. For example, "Either Ralph is a dog or Howie is not a cat, and George is a duck, but Birta is a dog" corresponds to (A v ~B) & (C & D). Using these determinations, we can model whether one claim follows as conclusion from a collection of other claims. Namely, it must be impossible for all the premises to be true and the conclusion false at the same time, which is the case based solely on the form of the propositions (relative to these connectives) if there is no assignment of truth-values to the premises that makes all of them true and the conclusion false.

This model or theory of reasoning is quite different from the other models we've seen. Though it is in some part descriptive and begins with abstractions from ordinary language, it is not the worse for finding that people do not reason as it describes. Though it may be thought of as describing nonsensible, abstract propositions of which it makes correct predictions, as a model of reasoning its role is *prescriptive*: it says that this is the correct way to reason.

Or rather, it says that this is the correct way to reason so long as the differences don't matter. But if the differences do matter, say if we wish to classify claims as true, false, or unknown, then this model won't be useful for drawing conclusions about how we should reason.

Typically in presenting this model, logicians do not point out the general principle on which it is based: the model is applicable when all that is of concern in reasoning is whether a claim is true or false and how it is built up in terms of these connectives. If we focus on the result of the abstracting and invoke abstract propositions as the subject matter of logic, then it is hard to see the general principle or even to conceive of the model as an abstraction from experience.

Example 9 Electrical switches—a new use of an old model
We can model circuits of electrical switches using the same tables as for classical logic. Instead of "true" and "false," T and F stand for "on" and "off," where these are the only two possibilities for the switches.

A series combination of switches A and B is:

•——(A)——(B)——•

The picture is meant to convey that there is a wire, the line, connecting the switches to two terminals, the dots. Current flows through a switch if it is on, and not through it if it is off. Current will flow through a series combination exactly when both switches are on, just as the table for & tells us.

A parallel combination of switches A and B is:

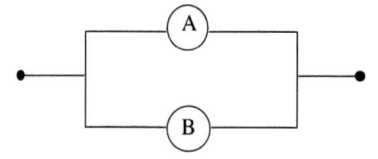

Current will flow through a parallel combination exactly when one or the other or both switches are on, just as the table for v tells us.

If we let ~A stand for a switch that is off when A is on, and on when A is off, then we can model any circuit built from these kinds of switches using the tables for &, v, and ~ . For example, (A v B) & ((~A & C) v (D & ~B)) models:

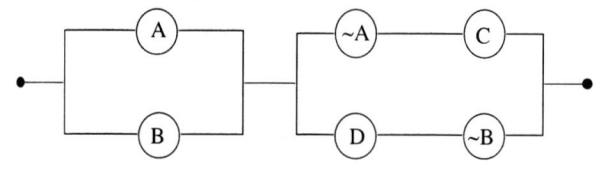

Depending on how we view the reasoning, the analogy is between the symbols and the way we manipulate them on the one hand and the circuits on the other, or it is between reasoning with claims and circuits. Either way, we can draw conclusions about the flow of electricity in very complicated circuits using this model, so long as we can ignore the differences: the wire has length and diameter and electrical resistance; switches aren't just on or off, but have some short period where the current is neither fully on or off when they are switched; and so on. This model is descriptive, not prescriptive.

Models, theories, and truth

We've seen models of static situations (the map) and of processes (acceleration of falling objects, gases in a container). We've seen examples of models that are entirely visual, intended as point-to-point comparisons, and of models formulated entirely in terms of mathematical equations. We have seen models where the assumptions of the model are entirely implicit (the map) and models where the assumptions are quite explicit (Newton's laws of motion).

In all the examples either the reasoning is reasoning by analogy or can be seen as proceeding by abstraction much as in reasoning by analogy.[4] We do not ask whether the assumptions of a theory or model are true, even if that was the intention of the person who created the theory. Rather, we ask whether we can use it in the given situation: Do the similarities that are being invoked hold and do the differences not matter? Even in the case of Newton's laws of motion, where it would seem that what is at stake is whether the assumptions are true, we continue to use the model when we know that the assumptions are false in those cases where, as in any analogy, the differences don't matter.[5] In only one example did it seem that what was at issue was whether a particular assumption of the theory was actually true of the world (the ether).

Laws in science are false when we consider them as representing all aspects of some particular part of our experience.[6] The key claim in every analogy is false in the same way. When we say that one side of an analogy is "just like" the other, that's false. What is true is that they are "like" one another in some key respects that allow us to deduce claims for the one from deducing claims for the other. In the same way we can use the assumptions of a theory so long as the differences—what we ignore—don't matter.

The exact conditions for an analogy or abstraction to qualify as a model or a theory are quite informal. The term *model* seems to be more readily applied to what can be visualized or made concrete, while the term *theory* seems to be applied to abstractions that are fairly formal and have very explicitly stated assumptions or general principles. But in many cases it is as appropriate to call an example a theory as to call it a model, and there seems to be no definite distinction between those terms.[7]

Theories and confirmation

From theories we can make predictions, and when a prediction turns out to be true we say that it confirms (to some degree) the theory. But this is not the same as confirming an explanation. Confirming an explanation is understood as showing (to some degree) that the claims doing the explaining are true, whereas we have seen that it rarely makes sense to say that the claims that make up a theory—the assumptions of the theory—are true or false.

Verifications of the relation of pressure, temperature, and volume in a gas do not confirm that molecules are hard little balls and that all collisions are completely elastic. A carpenter's square fitting exactly into a wooden triangle that is 5cm x 4cm x 3cm does not confirm the theorem of Pythagoras. Nor does finding a tree at the corner of 100 W and 100 S in Minersville disconfirm the model given by the map of Minersville.

Except in rare instances where we think (usually temporarily) that we have hit upon a truth of the universe to use as an assumption in a theory, we do not think that the assumptions of a theory are true or false. We can only say of a theory such as Euclidean plane geometry or the kinetic theory of gases that it is applicable or not in a particular situation we are investigating, where a "situation" is just some part of the world we describe using claims.

To say that a theory is *applicable* is to say that though there are differences between the world and what the assumptions of the theory state, those differences don't matter for the conclusions we wish to draw. Often we can decide if a theory is applicable only by attempting to apply it. We use the theory to draw conclusions in particular situations, claiming that the differences don't matter. If the conclusions — the predictions — turn out to be true (enough), then we have some confidence that we are right. If a prediction turns out false, then the model is not applicable there. We do not say that Euclidean plane geometry is false because it cannot be used to calculate the path of an airplane on the globe; we say that Euclidean plane geometry is inapplicable for calculating on globes.

When we make predictions and they are true, we confirm a range of application of a model. When we make predictions and they are false, we disconfirm a range of application; that is, we find limits for the range of application of a model. More information about where the model can be applied and where it cannot may lead, often with great effort, to our describing more precisely the range of application of a model. In that case, the claims describing the range of application can be added to the theory. We often use mathematics as a language for making the art of abstraction precise. For example, for Newton's laws of motion we can give limits on the size and speed of objects for which the theory is applicable. But for many theories it is difficult to state precisely the range of application. In other cases, such as the map of Minersville, it hardly seems worth the effort to state explicitly the range of application.

We want our theories to be as widely applicable as possible. Eventually we hope to find theories whose range of application can be precisely and clearly stated, where we can say that the theory is applicable whenever this is the case, where we are justified in saying that the theory is true.

But even then we would not be justified in saying that a particular claim that is used as an assumption of the theory is true. Rather, the claim is true in those situations in which the theory as a whole is applicable.[8]

Many other terms are used to describe theories: a theory is valid, a theory is true, a theory holds, a theory works for, ... I can find no other sense to these than to assimilate them to the question of whether a particular situation or class of situations to which we wish to apply a theory is within the range of the theory.

Modifying theories in the light of the evidence

How do we determine whether a theory is good? How do we determine whether one theory is better than another? We've seen that the criteria cannot in general include whether the assumptions of the theory are true, or, as is sometimes said, "realistic". Here is what Milton Friedman says in "The Methodology of Positive Economics":

> In so far as a theory can be said to have "assumptions" at all, and insofar as their "realism" can be judged independently of predictions, the relation between the significance of a theory and the "realism" of its "assumptions" is almost the opposite of that suggested by the view under criticism. Truly important and significant hypotheses will be found to have "assumptions" that are wildly inaccurate descriptive representations of reality, and, in general, the more significant the theory, the more unrealistic the assumptions (in this sense). The reason is simple. A hypothesis is important if it "explains" much by little, that is, if it abstracts the common and crucial elements from the mass of complex and detailed circumstances surrounding the phenomena to be explained and permits valid predictions on the basis of them alone. To be important, therefore, a hypothesis must be descriptively false in its assumptions; it takes account of, and accounts for, none of the many other attendant circumstances, since its very success shows them to be irrelevant for the phenomena to be explained.
>
> To put the point less paradoxically, the relevant question to ask about the "assumptions" of a theory is not whether they are

descriptively "realistic," for they never are, but whether they are sufficiently good approximations for the purpose in hand. And this question can only be answered by seeing whether the theory works, which means it yields sufficiently accurate predictions. pp. 14–15

Friedman, then, agrees with what I have said—until the last sentence. He and many others say that the sole criterion for judging whether a theory is good (or as he says, "valid") is whether it yields sufficiently accurate predictions. In that same paper he says:

> The only relevant test of the validity of a hypothesis is comparison of its predictions with experience. p. 8

Certainly it is important to get good predictions.[9] But if the assumptions are neither true nor true for the situation being analyzed, on what basis should we base our acceptance of further predictions? A good track record in the past? But who has been keeping score? Perhaps it is just judicious uses of the theory, always supplemented with assumptions—often unstated—that result in predictions that are sufficiently accurate. What allows us to distinguish between a theory that makes good predictions whose assumptions are clearly false on one hand and astrology on the other? After all, for many centuries astrology was the best theory around for divining human fate. Its predictions often came true, since they were sufficiently vague to allow that, and few people were keeping track of the predictions that turned out false.

There must be some criterion beyond the truth of predictions made using the theory that counts for whether a theory is good. Consider what we do when we discover that a prediction made using a theory is false.

When the theory of switching circuits predicts inaccurately for transistors, we look for what differences matter: what have we ignored in the case at hand that results in false predictions? Switches are not instantaneously on or off, there is electrical resistance, ... We then factor those aspects of the situation into our new theory.

When Newton's laws of motion result in inaccurate predictions for objects moving near the speed of light, we note that the theory had been assumed true for all sizes and speeds of objects and then restrict the range of application.

Even the prescriptive theory of classical propositional logic has been modified. When predictions made using it (certain inferences shown to be valid based on its assumptions) were found to be anomalous, counterintuitive, not good prescriptions for reasoning, attention turned to what aspects of claims had been ignored. Certain differences matter, and depending on what aspects are considered significant, such as subject matter or ways in which the claim could be true, different propositional logics have been set out as models of reasoning. [10]

On the other hand, when the theory of the ether resulted in false predictions, no modification was made to the theory, for none could be made. That theory did not abstract from experience, ignoring some aspects of situations under consideration, but postulated something in addition to our experience, something we were able to show did not exist. The theory was completely abandoned.[11]

When we find that a prediction is false, we have several choices:

1. The theory or model can be understood as an analogy or for use as reasoning by abstraction. That is, many aspects of our experience are ignored and only some few are considered significant. Then tracing back along that path of abstraction we can try to distinguish what difference there is between our model and our experience that matters. What have we ignored that cannot in this situation be ignored?

If we cannot state precisely what difference it is that matters in some general way, then at best the false prediction sets some limit on the range of applicability of the model or theory. We cannot use the theory here—where "here" means this situation or ones that we can see are very similar. If too many situations are eliminated, in particular ones that we used as archetypes in motivating the abstracting, we abandon the theory.

But we are driven to find precisely the difference that matters and try to factor it into our theory. That is, we try to devise a complication of our theory in which that aspect of our experience is taken into account. As with Einstein's improvement of Newton's laws, we get a better theory that is more widely applicable and that explains why the old theory worked as well as it did and why it failed in the ways it failed. We improve the map: by adding further assumptions we can pay attention to more in our experience, and that accounts for the differences between the theories.

2. Some theories, such as the theory of the ether, are not based on abstraction but postulate entities or aspects of the world in addition to what we have from our experience or other trusted theories. In that case a false prediction, or more usually many false predictions, lead us to consider such a postulate to be false. We abandon the theory.

3. Still, there are cases where we retain the theory even though it clearly contradicts experience. Astronomers tell us that the earth revolves around the sun, yet we see the sun travel across the sky while we are standing still. Physicists tell us that a table is mostly empty space, yet we knock on it and know it is solid. In these cases we have another alternative: We draw further consequences from the theory and use those along with observations about our perceptual capabilities to explain why such anomalous claims are not in contradiction with the theory.

4. One of our examples, however, does seem to fit Friedman's method for deciding whether a theory is good. Modern physics says that there is no preferred frame of reference in the universe. Hence, so far as the truth of the assumptions or their applicability to the case at hand, the Ptolemaic model of the universe is as good as the Copernican. But we choose the Copernican model because it yields better predictions. We can always modify the Ptolemaic system to account for observations by adding more epicycles. That will yield true predictions in some limited cases but will rarely work for objects other than the one for which an epicycle is posited. Moreover, the predictions we can make of the movement of planets and other objects in the solar system are clearer and simpler in the Copernican model.

To summarize:

- We create theories by abstracting from our experience.
- True predictions from such a theory confirm a range of application of the theory.
- False consequences from such a theory can be accomodated by either:
    —Modifying the theory to account for more of our experience.

- Restricting the range of application of the theory.
- Explaining the anomaly in terms of human perceptual capabilities.

• False predictions that cannot be accomodated in those ways lead us to abandon the theory. Either the resulting range of application is too small, or another theory works better for a wider range of application, or else we see that we have assumed something in addition to our experience that the false predictions lead us to believe is false.

The truth of the assumptions does matter for some theories: those based on claims that are not abstractions from experience. When it makes sense to talk of the truth or falsity of a particular dubious claim among the assumptions of a theory, deriving true predictions from the theory can help to confirm that claim in the same way we confirm an explanation; a false prediction may serve to prove it false.

A theory that is meant as an analogy or abstraction from our experience, ignoring much of what we wish to study and focusing only on what we consider the most important aspects, comes with an (often implicit) range of application. Deriving a true prediction from such a theory confirms to some extent that range of application. A false prediction can serve to set limits on the range of application. False predictions can also lead us to prefer a simpler theory based on abstraction that yields better predictions.[12]

True predictions are never enough to justify a theory. Indeed, the problem is that we do not "justify" a theory nor show that it is "valid." What we do in the process of testing predictions is show how and where the theory can be applied. For us to have confidence in that, either we must show that the claims in the theory are true or show in what situations the differences between what is represented and the abstraction of it in the theory do not matter. True (enough) predictions help in that. But equally crucial is our ability to trace the path of abstraction so that we can see what has been ignored in our reasoning and why true predictions serve to justify our ignoring those aspects of experience. Without that clear path of abstraction, all we can do is try to prove that the claims in the theory are actually true. Without that clear path or without reason to believe the claims are true, we have no more reason to trust the predictions of a theory than we have to trust the predictions of astrology.

Case study : A model with no path of abstraction

Milton Friedman's analysis of theories discussed above was meant to justify theories of conventional (classical, orthodox, neo-classical) economics, such as the theory of competitive equilibrium. Such theories begin with the assumption that all persons involved in the market are *rational*:

1. They are motivated solely by the goal of maximizing utility.
2. They have fixed, transitive preferences.
3. They have available to them all the relevant information and use it.
4. They reason perfectly.

These assumptions about people are false and have been demonstrated to be false in many behavioral studies.[13] This has led to the charge that predictions from such theories cannot be relied on. The theories are charged with having "unrealistic" premises.[14]

But as we saw above, and as Friedman stressed, false, unrealistic premises do not disqualify a theory from giving us good reason to believe that its predictions are true. From assumptions that are abstractions from experience and which in no sense could be true, not even in particular cases as with Euclidean geometry, we can reason to conclusions in which we are justified in having great confidence. Hence, the charge by many economists that theories based on the assumption that people are rational are unrealistic is not a serious criticism of such theories.

The question for such theories is what is the range of application for reasoning by them. When we obtain false predictions, how can we use those to clarify the limits of the range of application and improve the theory? And certainly with theories based on the assumption that people are rational, there are plenty of false predictions.[15]

Let's ask how we can trace back the path of abstraction for assumptions (1)–(4) to modify them to obtain better theories.

With (1) it's clear: we can assume other motives of people. Doing so complicates the models but can yield better theories. We can also assume that people want only to satisfy some level of utility rather than maximizing.[16]

But for each of (2)–(4) there is a problem: What is being abstracted from experience to create these assumptions of the theory?

Certainly people do not have fixed transitive preferences.[17] But what is it that we are ignoring in their behavior that allows us to assume they do? At best we can say that sometimes they have fixed transitive preferences, and in those cases the theory should apply.

When do people have all the relevant information and use it in making decisions? Perhaps they do in very restricted situations, such as buying and selling currencies. False predictions from this assumption would limit the theories to such cases. Or models can be devised that take into account limited access to relevant information.[18]

But for (4) there is no such obvious route to limiting the range of application of the theories. It is not simply that people do not normally reason well. It is not that they do not want to reason well. Rather, most people do not have the skills to reason as well as demanded by theories based on the assumption that people are rational, as any teacher of critical thinking can attest.[19] The norms of reasoning well are prescriptive, not descriptive.

There is no abstraction from experience in postulating that people reason perfectly. To assume that people are rational is to ascribe capabilities to people that they do not have. It is not like Euclidean geometry where we ignore much from experience; it is like postulating the existence of an ether, where we assume something of the world in addition to what we know there is.

There is no point along the path of abstraction where we can modify theories based on the assumption that people are rational in order to take into account some further aspect of their reasoning. The moment we assume that people reason well only some times, there is no longer a model. As with theories built on the assumption that there is an ether, when we encounter false predictions we can only abandon the theory.

But sometimes theories of conventional economics do predict well. Friedman would account for that and justify the use of such theories by saying that though people do not act rationally, they act "as if" they are rational. It is worth quoting from that same paper at length to see how his justification proceeds.

> Consider the density of leaves around a tree. I suggest the hypothesis that the leaves are positioned as if each leaf deliberately sought to maximize the amount of sunlight it receives, given the position of its neighbors, as if it knew the physical laws determining the amount of

sunlight that would be received in various positions and could move rapidly or instantaneously from any one position to any other desired and unoccupied position. Now some of the more obvious implications of this hypothesis are clearly consistent with experience: for example, leaves are in general denser on the south than on the north side of trees but, as the hypothesis implies, less so or not at all on the northern slope of a hill or when the south side of the trees is shaded in some other way. Is the hypothesis rendered unacceptable or invalid because, so far as we know, leaves do not "deliberate" or consciously "seek," have not been to school and learned the relevant laws of science or the mathematics to calculate the "optimum" position, and cannot move from position to position? Clearly, none of these contradictions of the hypothesis is vitally relevant; the phenomena involved are not within the "class of phenomena the hypothesis is designed to explain"; the hypothesis does not assert that leaves do these things but only that their density is the same *as if* they did. Despite the apparent falsity of the "assumptions" of the hypothesis, it has great plausibility because of the conformity of its implications with observation. We are inclined to "explain" its validity on the ground that sunlight contributes to the growth of leaves and that hence leaves will grow denser or more putative leaves survive where there is more sun, so the result achieved by purely passive adaptation to external circumstances is the same as the result that would be achieved by deliberate accommodation to them. This alternative hypothesis is more attractive than the constructed hypothesis not because its "assumptions" are more "realistic" but rather because it is part of a more general theory that applies to a wider variety of phenomena, of which the position of leaves around a tree is a special case, has more implications capable of being contradicted, and has failed to be contradicted under a wider variety of circumstances.[20]

Friedman's hypothesis about leaves seeking to maximize the amount of sunlight they receive cannot be used for reasoning by analogy or abstraction. It does not begin by either (a) looking at a real or at least possible situation and comparing it to the growth of leaves, allowing us to distinguish what are the similarities and what are the differences, or (b) abstracting from experience to state exactly what are the points of similarity that are supposed to hold, ignoring all else.

Rather, what he has posited is not an abstraction but the addition of properties to a given situation. We are asked to suppose that leaves behave anthropomorphically with the skills of a terrific calculator. And then we are asked to ignore that as well. This doesn't make sense as a method of reasoning: why should we have confidence that predictions made from such a hypothesis will be accurate? That some of the predictions turn out to be accurate cannot be enough, any more than they are in astrology. We need to know why they turn out accurate in order to have confidence in the theory or model.

The alternative hypothesis of passive adaptation that he presents is better, but not for the reasons he gives; rather, it is better for the reason he says is not meaningful. Namely, no clearly false assumptions incapable of fitting into reasoning by analogy or abstraction have been made.

Friedman gives another example, which leads to the rationality assumption:

> ... the hypothesis that the billiard player made his shots *as if* he knew the complicated mathematical formulas that would give the optimum directions of travel, could estimate accurately by eye the angles, etc., ... Our confidence in this hypothesis is not based on the belief that billiard players, even expert ones, can or do go through the process described; it derives rather from the belief that, unless in some way or other they were capable of reaching essentially the same result, they would not in fact be *expert* billiard players.
>
> It is only a short step from these examples to the economic hypothesis that under a wide range of circumstances individual firms behave *as if* they were seeking rationally to maximize their expected returns (generally if misleadingly called "profits") and had full knowledge of the data needed to succeed in this attempt; *as if*, that is, they knew the relevant cost and demand functions, calculated marginal cost and marginal revenue from all actions open to them, and pushed each line of action to the point at which the relevant marginal cost and marginal revenue were equal. Now, of course, businessmen do not actually and literally solve the system of simultaneous equations in terms of which the mathematical economist finds it convenient to express this hypothesis, any more than leaves or billiard players explicitly go through complicated mathematical calculations or falling bodies decide to create a vacuum. ...
>
> Confidence in the maximization-of-return hypothesis is justified by evidence of a very different character [from the truth of its assumptions]. This evidence is in part similar to that adduced on behalf of the billiard-player hypothesis—unless the behavior of businessmen in some way or other approximated behavior consistent

42 *Reasoning in Science and Mathematics*

with the maximization of returns, it seems unlikely that they would remain in business for long. Let the apparent immediate determinant of business behavior be anything at all—habitual reaction, random chance, or whatnot. Whenever this determinant happens to lead to behavior consistent with rational and informed maximization of returns, the business will prosper and acquire resources with which to expand; whenever it does not, the business will tend to lose resources and can be kept in existence only by the addition of resources from the outside. The process of "natural selection" thus helps to validate the hypothesis—or, rather, given natural selection, acceptance of the hypothesis can be based largely on the judgment that it summarizes appropriately the conditions for survival.[21]

Just as Friedman is wrong about the leaves, he is wrong about the billiard player and hence wrong in his justification of the maximization-of-return hypothesis. Moreover, he has hit—unwittingly—on the greater problem with both the billiard player and the maximization-of-return hypotheses: they can only explain success, not failure of the agent. They apply to much less of the situation than is needed to make a good model: How do we explain the failure of a billiard player to make a shot—he was calculating badly? How do we explain the bankruptcy of a company—they weren't calculating correctly? A good model of either situation has to account for failure as well as success, for all the "players."

Friedman introduces his "as if" talk in order to justify economic theories based on assuming that people are rational. The problem with the rationality assumptions, as in his examples, is that they are not abstractions but positing of properties that are clearly false. It matters how assumptions are derived from experience in order for us to be justified in saying that they do not have to be true to get good predictions. If they are obtained by abstraction, then they need not be true: all that matters is the scope of their application. We can trace back the path of abstraction. But with hypotheses that postulate additional properties, if those postulates are not true—that is, if the model is not applicable in the case at hand—we have no reason to trust the model.[22]

It will be difficult to give criteria for what we mean when we say that an assumption is not an abstraction from experience but rather postulates something of the world in addition to our experience. But those who wish to base theories on assumptions that are clearly false have the burden to show in what way their theories arise by abstraction; it is for them to justify their theories.

Economic theories have been developed in the last few decades that abandon the assumption that people are rational.[23] They are much more complicated. They rarely yield predictions of specific events so much as describe overall behavior of an economy, presenting something like the normal conditions from which causal claims could be derived. They are to conventional economics much as applied mathematics is to pure mathematics. As with pure mathematics, conventional economics has a great beauty and intellectual attraction and internal coherence that make it of interest to many and which justify its study. It is up to those who practice it, however, to justify further why we should have any confidence in its predictions as applied to experience.[24] I hope to have shown here on what grounds we can judge such justifications.

Case study: Confusing applicability with truth

Did you wash your face with soap and water?

Did you just drink a glass of water?

We're asked questions like these every day. We can answer them because we know what the word "water" means. So it must be a discovery, some new fact about the world, when chemists tell us:

(1) Water is H_2O.

Of course we have to understand this as shorthand. After all, "H_2O" is a description of a kind of molecule, and water is a substance. The claim must really be:

(2) Every quantity of water is a collection of H_2O molecules.

But if this is right, we have to reject almost every way we have of learning what "water" means. We point to a river and say "That's water," yet it isn't a collection of H_2O molecules: it's a mixture, with sand, and dirt, and minerals. We never encounter collections of just H_2O molecules. No one, on this account, has ever seen or touched or drunk water, except perhaps for a few chemists in a laboratory.[25]

Since (1) is supposed to be an identity, the reverse of (2) is part of it, too:

44   *Reasoning in Science and Mathematics*   false

(3) Every collection of H$_2$O molecules is a quantity of water.

But to accept this we have to reject almost every claim we believe is true about water: "Water flows," "Water freezes," "Water gives rise to a sensation of wetness," "Water evaporates." None of these are true of a single molecule, not even of four molecules of water. To rescue those claims we have to modify both (2) and (3) to give a lower limit for the number of molecules of H$_2$O that can count as a quantity of water.[26] Perhaps such a number could be established; to do so would be to substitute a precise notion for an imprecise one, avoiding vagueness, though at the price of some anomalous classifications.

Yet we do talk of water as if it were the stuff that is basic in many liquids we encounter. We talk of dirty water, alkaline water, acidic water, salt water. But compare: a brown dog, a white dog, a tiny dog. Just as each of those is a dog, so, too, alkaline water is water. Still, we talk of pure water as if we could eliminate what's not really water in a muddy river or a cup of salt water, while we don't talk of a pure dog. This is the first step in science: we take some part of our experience and choose to ignore all but one aspect. We say that dissolved minerals and suspended bits of earth are not what we're interested in, they are not important to our discussion of this liquid.

To assume that there was a pure substance which was the substratum, the key ingredient in all the liquids we call "water," was at one time only a hope, a conjecture. Chemists have now isolated that key ingredient. One chemist told a colleague of mine who asked him if water is H$_2$O:

> To answer the question simply, water is H$_2$O. In general, a material is defined by its chemical formula and an atom is defined by the number of protons it contains.

But this takes "Water is H$_2$O" to be a definition. No one, then, could have discovered that water is H$_2$O.

I told my colleague to go to the chemistry professor, take a glass and fill it with water from a tap in front of him, and ask him if that's water. If he says "yes," he's wrong by his definition. If he says "no" then ask him what good are his theories and analysis of water as H$_2$O.

"Water is H$_2$O" is not an identity, but it is meaningful. It shows us how to use chemical and physical theories. We can use those theories to analyze water and how water is in the world so long as what those theories take into account is what we want to take into account and so long as what those theories ignore is what we are willing to ignore. A chemical theory will not say "water freezes" because "water" is not a term of the theory. It will say that given a large enough quantity of H_2O molecules, and only H_2O molecules, they will freeze at 0 degrees centigrade. We then use that information to find our way in the world. Actually, we knew long before there were any theories of molecules that water—certainly not "pure" water—freezes at some temperature that we designate as 0 degrees centigrade. It was for the chemical theory to explain that. And it does, if we agree to use the theory correctly in applying it to experience: we only use it to talk about water if we have a large enough quantity of H_2O molecules with few enough "impurities."

The claim "Water is H_2O" is shorthand, a direction for how to use our chemical theories in analyzing our experience.[27]

Notes

1. (p. 20) But when I asked a friend from Colombia to read a map when we were driving, she was incapable as she had no experience reading road maps.

    See Alberto Cordero in "The Infinitely Faceted World: Intimations from the 1950s" for a discussion that illuminates the points I make here.

2. (p. 26) The role of mathematics in science is explored in the second appendix of the essay "Mathematics as the Art of Abstraction" in this volume.

3. (p. 27) The relation between Newton's theory and modern physical theories is often characterized in terms of a correspondence. Robert T. Weidner and Robert L. Sells in *Elementary Modern Physics* say:

    > Any theory or law in physics is, to a greater or lesser degree, tentative and approximate. This is true because applying a physical law to situations to which it has not been experimentally tested *may* show it to be incomplete or even incorrect. Thus, when we extrapolate a theory to untested situations, we cannot be sure that the theory will hold. However, if a new, more general theory is proposed, there is a completely reliable guide for relating this more general theory to the older more restricted theory. This guide is the *correspondence principle*, first applied to the theory of atomic structure by the Danish physicist Niels Bohr in 1923. We shall find it helpful to apply this principle in a broadened sense, using it to great advantage in relativity physics as well as in quantum physics.
    >
    > THE CORRESPONDENCE PRINCIPLE: *We know in advance that any new theory in physics—whatever its character or details—must reduce to the well-established classical theory to which it corresponds when the new theory is applied to the circumstances for which the less general theory is known to hold.*
    >
    > Consider a simple, familiar situation that illustrates the correspondence principle. When we have a problem in projectile motion of relatively small range, the following assumptions are made: (1) the weight of the projectile is constant in magnitude, given by the mass times a *constant* gravitational acceleration, and (2) the Earth can be represented by a plane surface, and (3) the weight of the projectile is constant in direction, vertically downward. With these assumptions, a parabolic path is predicted and one gets perfectly satisfactory results provided that the projectile motion extends over only relatively short distances. But if we try to describe the motion of an Earth satellite with the same assumptions, *very* serious errors will be made. To discuss the satellite motion one must assume, instead, that (1) the

weight of the body is *not* constant but varies inversely with the square of its distance from the Earth's center, (2) the Earth's surface is round, not flat, and (3) the direction of the weight is *not* constant but always points toward the Earth's center. *These* assumptions lead to a prediction of an elliptical path, and to a proper description of satellite motion. Now if we apply the second, more general theory to the motion of a body traveling a distance small compared to the Earth's radius at the surface of the Earth, notice what happens: the weight appears to be constant both in magnitude and direction, the Earth appears flat, and the elliptic path becomes parabolic. This is precisely what the correspondence principle requires!

The correspondence principle asserts that when the conditions of the new and old theories correspond, the predictions will also correspond. We have then an infallible guide with which to test any new theory or law: it must reduce to the theory which it supplants. Any new proposed theory which fails to meet this test is clearly defective in so fundamental a way that it cannot possibly be accepted.

p. 29 [italics in original]

This is a confused way of saying that if we choose to ignore certain differences, then the two theories will give similar predictions. Newton's theory and general relativity, for example, never give exactly the same predictions. Moreover, we don't want our current theory of chemical combustion to "reduce" to the phlogiston theory when the phlogiston theory predicted correctly.

4. (p. 31) Mary B. Hesse in *Models and Analogies in Science* defends the view that analogies are essential for the development and interpretation of scientific theories. But in doing so she shows the limitation of the use of analogy if not extended to the more general process of abstraction.

5. (p. 31) In their 1965 textbook *Mechanics* C. Kittel, W. D. Knight, and M. A. Ruderman say:

> *Newton's third law.* Whenever two bodies interact, the force F_{12} on the second body (2) due to the first body (1) is equal and opposite to the force F_{21} on the first (1) due to the second (2): $F_{12} = -F_{21}$. There are inherent limitations to the validity of the third law: we believe . . . that all signals or forces have a finite propagation velocity. The third law, however, states that F_{12} is equal and opposite to F_{21} when both are measured *at the same time*. This requirement is inconsistent with the finite time interval required for one particle to feel the force the second particle is exerting. In atomic collisions the third law is therefore not always a good approximation. In automobile collisions it is quite a good approximation because the duration of the collision is

long in comparison with the time it takes a light signal to traverse a crumpled automobile. pp. 55–56

6. (p. 31) Michael Scriven discusses this in "The Key Property of Physical Laws—Inaccuracy" and "Explanations, Predictions, and Laws". In the latter he says:

> The examples of physical laws with which we are all familiar are distinguished by one feature of particular interest for the traditional analyses [of explanations]—they are virtually all known to be in error. Nor is the error trifling, nor is an amended law available which corrects for all the error. The important feature of laws cannot be their literal truth. p. 212

Nancy Cartwright in "The Reality Of Causes in a World of Instrumental Laws" says:

> What is important to realize is that if the theory is to have considerable explanatory power, most of its fundamental claims will not state truths, and that this will in general include the bulk of our most highly prized laws and equations. p. 381

7. (p. 31) Jack Birner in *The Cambridge Controversies in Capital Theory* says:

> It is quite common to use "theory" for a set of abstract propositions and "model" for a set of propositions on a lower level of abstraction. ... [T]he question of whether or not something is called a theory or a model is mostly a matter of convention. What matters is how theories or models of different degrees of idealization are related, i.e. their relative levels of abstraction. Therefore, I will use "theory" and "model" interchangeably.

8. (p. 33)

> Galileo said that a body subject to no forces has a constant velocity. (This is called Newton's first law of motion.) We have seen that this statement is true only in an inertial reference system—it defines an inertial system.   Kittel, Knight, and Ruderman, *Mechanics*, p. 61

9. (p. 34) Compare Stephen E. Landsburg, *The Armchair Economist*, p. 10, "Assumptions are not tested by their literal truth but by the quality of their implications."

10. (p. 35) See my *Propositional Logics*, a summary overview of which is given in "A General Framework for Semantics for Propositional Logics" in *Reasoning and Formal Logic* in this series.

11. (p. 35) David Isles (personal communication) disagrees: "How is the ether any less of an abstraction from experience than, say, the notion of force or even the notion of energy as it appears in physics? Thus, we experience water which carries water waves, we experience less substantial air which carries sound waves, and (a slight abstraction from experience) we have a very insubstantial fluid which carries light waves." Isles is correct that we can hypothesize the existence of an ether by analogy with those other mediums of transmission, but that is distinctly different from beginning with some part of experience and ignoring some of its properties. The notions of force and energy in physics are not postulated to be things or substances; see the discussion of this point in relation to the nature of cause and effect in "Reasoning about Cause and Effect" in *Cause and Effect, Conditionals, and Explanations* in this series.

12. (p. 37) There are great difficulties in explaining what we mean by one theory being simpler than another. Philipp G. Frank outlines those in "The Variety of Reasons for Acceptance of Scientific Theories." One example he gives is:

> There was a time when, in physics, laws that could be expressed without differential equations were preferred, and in the long struggle between the corpuscular and the wave theories of light, the argument was rife that the corpuscular theory was mathematically simpler, while the wave theory required the solution of boundary problems of partial differential equations, a highly complex matter. We note that even a purely mathematical estimation of simplicity depends upon the state of culture of a certain period. People who have grown up in a mathematical atmosphere—that is, saturated with ideas about invariants—will find that Einstein's theory of gravitation is of incredible beauty and simplicity; but to people for whom ordinary calculus is the center of interest, Einstein's theory will be of immense complexity, and this low degree of simplicity will not be compensated by a great number of observed facts.
>
> However, the situation becomes much more complex, if we mean by *simplicity* not only simplicity of the mathematical scheme, but also simplicity of the whole discourse by which the theory is formulated. We may start from the most familiar instance, the decision between the Copernican (heliocentric) and the Ptolemaic (geocentric) theories. Both parties, the Roman Church and the followers of Copernicus, agreed that Copernicus' system, from the purely mathematical angle, was simpler than Ptolemy's. In the first one, the orbits of planets were plotted as a system of concentric circles with the sun as center, whereas in the geocentric system, the planetary orbits were sequences of loops. The observed facts covered

50 Reasoning in Science and Mathematics

by these systems were approximately the same ones. The criterions of acceptance that are applied in the community of scientists today are, according to the usual way of speaking, in agreement with observed facts and mathematical simplicity. According to them, the Copernican system had to be accepted unhesitatingly. Since this acceptance did not happen before a long period of doubt, we see clearly that the criterions "agreement with observed facts" and "mathematical simplicity" were not the only criterions that were considered as reasons for the acceptance of a theory. pp. 14–15

Those who invoke simplicity as a criterion for choosing one theory over another have much to do to make that idea sufficiently precise to discriminate in an acceptable manner between competing theories.

13. (p. 38) See Richard H. Thaler, *The Winner's Curse*.

14. (p. 38) For a discussion and references see Mark Blaug, "Economic Methodology in One Easy Lesson."

15. (p. 38) See, for example, Blaug, "Economic Methodology in One Easy Lesson," or Thaler, *The Winner's Curse*, or Paul Ormerod, *Butterfly Economics*.

16. (p. 38) See Gerd Gigerenzer and Reinhard Selzen "Rethinking Rationality," which also discusses the history of this notion of rationality.

17. (p. 39) Thaler in *The Winner's Curse* cites numerous psychological studies to this effect.

18. (p. 39) See again Gigerenzer and Selten, "Rethinking rationality."

19. (p. 39) Moreover, the assumption of rationality requires that people can survey all information and all consequences of certain assumptions at once. That is, people are assumed to be able to survey a completed infinite set. This contrasts with, say, Euclidean geometry, which, in applications, requires only that we can always find an additional point as required by the axioms.

20. (p. 40) "The Methodology of Positive Economics," pp. 19–20.

21. (p. 42) "The Methodology of Positive Economics," pp. 21–22.

22. (p. 42) There is another oddity in saying that people act as if they are rational when indeed they are not. Typically, the evidence we invoke for claiming a person is rational is how he or she acts, as discussed in "Rationality" in *Prescriptive Reasoning* in this series. See the articles in *Fictions in Science* edited by Mauricio Suárez for attempts to justify "as if" talk.

23. (p. 43) See Ormerod, *Butterfly Economics* for a non-technical introduction.

24. (p. 43) Similar comments apply to decision theory as it is customarily formulated, for it, too, assumes people to be perfect reasoners. See, for example, Michael Resnik, *Choices*. There is a similar debate about how to judge theories in modern physics. Richard P. Feynman in *QED*, p.10, represents one view that is similar to Milton Friedman's:

> [Physicists have] learned that whether they like a theory or they don't like a theory is *not* the essential question. Rather, it is whether or not the theory gives predictions that agree with experiment. It is not a question of whether a theory is philosophically delightful, or easy to understand, or perfectly reasonable from the point of view of common sense. The theory of quantum electrodynamics describes Nature as absurd from the point of view of common sense. And it agrees fully with experiment. So I hope you can accept Nature as She is—absurd.

Others, particularly Albert Einstein, argue that quantum mechanics is incomplete precisely because all it offers is good predictions but not—in the terms described here—a clear path of abstraction. See N. P. Landsman, "When Champions Meet: Rethinking the Bohr-Einstein Debate."

25. (p. 43) And there's good reason to think none of them have either, for obtaining a sample of H_2O molecules large enough to drink that contains no isotopes or ions is virtually impossible.

26. (p. 44) "Cluster physics" is the study of the minimal number of molecules of a particular kind that is necessary for a collection of those molecules to have the macroscopic properties we associate with the substance.

27. (p. 45) Paul Needham in "Micoressentialism: What is the Argument?" criticizes and surveys problems with the identification of water with H_2O.

Experiments

Experiments in science are meant to produce observations that confirm or disconfirm a theory or to lead to new conjectures. Looking at examples of experiments we can get a better idea of what scientists expect of a good experiment.

Observational claims 53
Examples
    1. A recipe from a famous coffee house 55
    2. Feeding behavior of primates 56
    3. Cyclic variations in grass growth 57
    4. The refraction of light rays 58
    5. Measurement of photon bunching in a thermal light beam 59
    6. Testing for anomalous cognition (ESP) 60
    7. The growth of living nerve cells in vitro 61
Experiments as the basis of correlations for causal reasoning . . 62

Observational claims

Observational claim An observational claim is one established by personal experience or observation in an experiment.

Evidence Evidence is usually the observational claims used as premises of an argument. Sometimes the term refers to all the premises.

What do we mean by "observation in an experiment"?
A physicist may say she saw an atom traverse a cloud chamber, when what she actually saw was a line made on a piece of photographic film. A biologist may say he saw the nucleus of a cell, when what he saw was an image projected through a microscope. In both cases they are not reporting on direct personal experience but on deductions made from their experience. However, those claims made by deduction from the perceptions arising from certain types of experiments are, by consensus in that area of science, deemed to be observations.

54   *Reasoning in Science and Mathematics*

Within any one area of science there is a high level of agreement on what counts as an observational claim. But from one area of science to another that standard may vary. A physicist beginning work in biology may well question why certain claims are taken as obvious deductions from experience, such as the reality of what you see through a microscope. But after the general form of the inference is made explicit once or twice—from such direct claims about personal experience to the observational claims—the physicist is likely to accept such claims as undisputed evidence. If he doesn't accept such deductions, he is questioning the basis of that science.

When new techniques are introduced into a science or a new area of science is developing, there is often controversy about what counts as an observational claim. Galileo's report of moons around Jupiter was received with considerable skepticism because telescopes were not assumed to be accurate, and indeed at that time they distorted a lot. In ethology, the study of animal behavior in natural settings, there is no agreement yet on what counts as an observational claim, and you can find different journal articles using different standards. For example, consider:

Some would describe this as an incident of the first chimpanzee getting angry and chasing the second one away and then the second returning to pacify and re-establish bonds with the first. That's what they saw. But others say that such a description is loaded with assumptions that have not been established, such as that chimpanzees have emotions sufficiently similar to humans to label as anger and that chimpanzees intend to accomplish certain ends rather than operating instinctually.

One constraint we impose on reports of observations is they should be replicable. We believe that nature is uniform. What can happen once can happen again, *if* the conditions are the same. Scientists typically won't accept reports on observations that they are unable to reproduce.

Duplicable and replicable experiments An experiment is *duplicable* means it is described clearly enough that others can follow the method to obtain observations. It is *replicable* if when it is duplicated the observations of the new experiment are in close agreement with the observations of the original experiment.

The difficulty is to specify exactly what conditions are required and what counts as close enough agreement. It's fairly easy in chemistry and physics; less so in biology; much more difficult in psychology or ethology. It's virtually impossible in history and economics, which means history and economics are not sciences, except to the extent that we can describe very general conditions that may recur.

Examples

Example 1   A recipe from a famous coffee house

    Vegetarian Chile
2 cans each (include liquid) of:
    Pinto beans
    Great Northern beans
    Red beans
    Kidney beans
    Chili beans
1 # 10 can diced Tomatoes
Garlic   6-8 cloves chopped
Bell Pepper   1 chopped
Jalapeño Peppers   3 chopped
Chili Powder   2 soup spoons
Onions   2 chopped or in food processor
Paprika   1 soup spoon

Put in soup tureen and heat to boil for 1 hour. Take care the beans don't stick to the bottom.

    Analysis Any expert in the subject (any person who has worked in a commercial kitchen) will know what a #10 can of tomatoes is. Though "chopped" and "soup spoon" may be unclear, anyone who saw the chile being made would be able to duplicate the preparation.

Example 2   Feeding behavior of primates
General Methodology
Data were collected simultaneously on both the activity of the animals and the forest strata at which this activity took place. Counts were made at five-minute intervals of the numbers of individuals engaged in each of the six activities and the level of the forest in which the activity was performed. The following activities were recorded: feeding—the animal actually in the process of ingesting or picking a food item; grooming—mutual and self-grooming were distinguished for certain analyses; resting—no body displacement, or feeding, or grooming, sunning, etc.; moving—movement of an individual, including individual foraging; travel—movement of the group; and other—e.g., sunning, play, fighting. These data were collected only after the animals under observation were reasonably habituated to the observer. Each observation of an animal constituted an individual activity record (IAR) collected in a given five-minute time sample. Because of the focus of the study and the difficulty in keeping continuous contact with an individual animal, no attempt was made to follow individual animals nor to collect statistical data on specific age or sex classes. Statistical analyses of the data were complicated by the fact that some of the activity records were not independent of each other. The methods used for the statistical analyses are reported in Sussman *et al*.

To determine levels of the forest, I used Richards' (1957) categories of forest stratification as a model and assigned numbers of one to five to the forest layers. Level 1 is the ground layer of the forest; it includes the herb and grass vegetation. Level 2 is the shrub layer, from one to three metres above the ground. This layer is usually found in patches throughout the continuous canopy forest, but is much more dense and is the dominant layer in the brush and scrub regions. Level 3 of the forest consists of small trees, the lower branches of larger trees, and saplings of the larger species of trees. This layer is about three to seven metres high. Level 4 is the continuous or closed canopy layer. It is about five to 15 metres high. The dominant tree of the closed canopy, at all three forests, is the kily (*Tamarindus indica*). Level 5 of the forest is the emergent layer and consists of the crowns of those trees which rise above the closed canopy. It is usually over 15 metres high.

All three forests in which I made intensive studies were primary forests and the tree layers were quite distinct. In most cases, the particular level in which an animal was observed could be distinguished easily. If I could not determine the forest level unambiguously, I did not record it.

Observations recorded in this manner may be biased because animals that are active in certain levels of the forest may be more difficult to see than those active at other levels. I attempted to minimize this problem by following a relatively small number of animals (usually from five to ten) throughout a period of continuous observation, keeping track of all the animals. For *Lemur fulvus* this usually included the whole group, which was small and, for the most part, moved together. It was more difficult to do this when observing *Lemur catta*, for which it was often necessary to follow and observe subgroups of the larger group. The larger group would disperse, especially during foraging and feeding, and during afternoon rest periods.

Day ranges were mapped by following a group from one night resting site in the morning to the time it settled in another night resting site in the evening. The location of the group was plotted throughout the day on a prepared map of the forest and the amount of time the group spent in each location was recorded. Home ranges include the sum of all the day ranges. The data on home ranges are limited, however, and probably do not represent total home ranges of the groups, since the study in each area was limited to a few months.

R.W. Sussman, "Feeding behaviour of Lemur Catta and Lemur Fulvus"

Analysis It is difficult to be more precise than this in ethology. The description of the methodology is clear enough to count as duplicable, perhaps even by someone who isn't an expert in the subject. Whether the observations are replicable will depend on how closely we expect them to agree with the ones in this paper.

Note that the author has not stated what time of year the observations were made nor the percentage of males versus females in the groups he studied. These are not part of the conditions that need to be duplicated; implicitly, the author is saying they don't matter. If it turns out in trying to duplicate this experiment that different observations are obtained at different times of the year, then the time of year would have to be added as part of the conditions that are important and which have to be duplicated.

Example 3 Cyclic variations in grass growth

Grass exhibits a cyclical growth pattern surprisingly different from any other known plant. In this study, average grass blade heights have been measured, on a daily basis, over a 10 week period. Measurements were taken, utilizing vernier calipers, of the height of one hundred individual grass blades randomly chosen in a 10 foot square area positioned in front of an apartment complex in the Lexington, Kentucky area. (Measurements were also repeated with a different set of calipers to ensure reproducibility on a different apparatus.) The average of these measurements was computed and experimental error was taken as the standard deviation of the mean divided by the square root of the number of grass blades in the average. The procedure was repeated on a daily basis for a period of 10 weeks.

Figure 1: Experimental measurements of average grass height are plotted versus time. Solid line represents experimental data. Short dashed line indicates a "constant grass height" calculation and is normalized to the experimental data to produce the best fit.

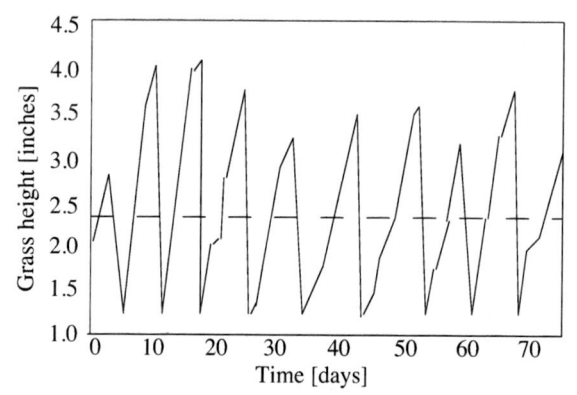

Results and Discussion The average grass heights, measured in this work, are plotted as a function of time in Figure 1. As one can readily see, there exists a periodic variation in average grass height with an approximate cycle of 7 to 10 days. Another intriguing observation is that there exists a minimum grass height, or "grass baseline," of about 1.3 inches.

Since the cyclic period of the grass is 7 to 10 days, one may conclude that grass height varies on a "week-about" basis. The physical mechanism responsible for this cyclic grass height phenomenon is not clearly understood at this time."   V. D. Irby, M. S. Irby, Dept. of Physics and Astronomy, University of Kentucky, *Annals of Improbable Research*, Vol. 1, no. 4, 1995.

Analysis This experiment was done to the highest standard of duplicability and indeed, we are sure, replicability, at least at the times specified. But, as you have surely noted by now, it is worthless, for it describes a common phenomenon that we can explain quite easily. What has the form of a good experiment need not be a good experiment.

Example 4 The refraction of light rays
In the wall or window of a room let F be some hole through which solar rays OF are transmitted, while other holes elsewhere have been carefully sealed so that no light enters from any other place. The darkening of the room, however, is not necessary; it only enables the experiment to turn out somewhat more clearly. Then place at that hole a triangular glass prism AaBbCk that refracts the rays OF transmitted through it toward $PYTZ$.

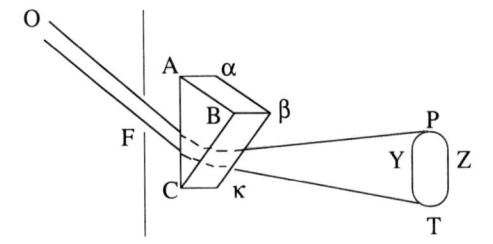

Isaac Newton, *Optica*, Part 1, Lecture 1, 1670.

Analysis This is very clear because of the diagram. It can be and often was duplicated, and the observations were replicated.

Example 5 *Measurement of photon bunching in a thermal light beam*

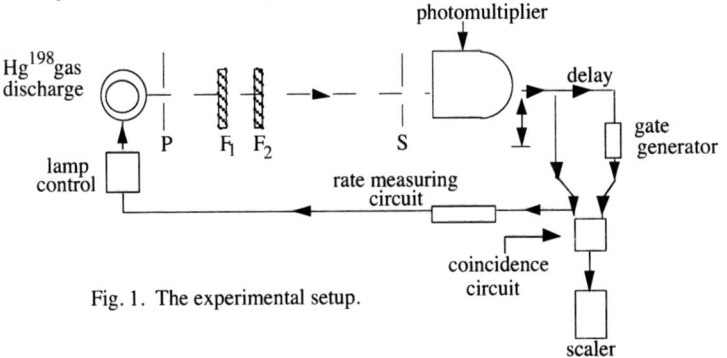

Fig. 1. The experimental setup.

The experimental setup ... is shown in Fig. 1. A light beam from a low-pressure Hg^{198} gas discharge lamp passed through a pin-hole P (diameter 0.54 mm), an optical filter F_1 that isolated the blue 5461-Å line, and a linear polarizer F_2. The beam fell on a 56 AVP photomultiplier through a rectangular aperture S (0.37 x 0.47 mm^2), whose dimensions were small enough to ensure a degree of coherence of at least 90% across the beam.

The pulses from the photomultiplier were shortened by reflection in a 1-nsec clipping line, and were fed to a specially designed gated pulse counter, which registered an output whenever two pulses appeared at the anode of the photomultiplier with a time separation lying between t_1 and t_2. The time t_1 was determined by the difference of two cable lengths, and could be varied by varying one of the lengths. t_1-t_2 was determined by the width of the gating pulse, and remained constant and equal to about 7.5 nsec as t_1 was varied.

B. L. Morgan and L. Mandel, *Phys. Rev. Lett.*, 1966.

Analysis If this is duplicable, you'd have to be an expert to follow the directions.

Example 6   Testing for anomalous cognition (ESP)
The vast majority of anomalous cognition experiments at SRI [Stanford Research Institute] and SAIC [Science Applications International Corporation] used a technique known as remote viewing. In these experiments, a viewer attempts to draw or describe (or both) a target location, photograph, object, or short video segment. All known channels for receiving the information are blocked. Sometimes the viewer is assisted by a monitor who asks the viewer questions; of course, in such cases the monitor is blind to the answer as well. Sometimes a sender is looking at the target during the session, but sometimes there is no sender. In most cases the viewer eventually receives feedback in which he or she learns the correct answer, thus making it difficult to rule out precognition [knowing the future] as the explanation for positive results, whether or not there was a sender.

Most anomalous cognition experiments at SRI and SAIC were of the free-response type, in which viewers were asked simply to describe the target. . . .

The SAIC remote-viewing experiments and all but the early ones at SRI used a statistical evaluation method known as rank-order judging. After the completion of a remote viewing, a judge who is blind to the true target (called a blind judge) is shown the response and five potential targets, one of which is the correct answer and the other four of which are "decoys." Before the experiment is conducted, each of those five choices must have had an equal chance of being selected as the actual target. The judge is asked to assign a rank to each of the possible targets, where a rank of 1 means it matches the response most closely, and a rank of 5 means it matches the least.

The rank of the correct target is the numerical score for that remote viewing. By chance alone the actual target would receive each of the five ranks with equal likelihood, since, despite what the response said, the target matching it best would have the same chance of selection as the one matching it second best and so on. The average rank by chance would be 3. Evidence for anomalous cognition occurs when the average rank over a series of trials is significantly lower than 3. (Notice that a rank of 1 is the best possible score for each viewing.)

This scoring method is conservative in the sense that it gives no extra credit for an excellent match. A response that describes the target almost perfectly will achieve the same rank of 1 as a response that contains only enough information to pick the target as the best choice out of the five possible choices.   Jessica Utts, "An assessment of the evidence for psychic functioning"
The Journal of Parapsychology, vol. 59, n. 4, p. 289, 1995.

Analysis   What does "All known channels for receiving information are blocked" mean? We need to know the exact layout of the room where the experiment was done. "In most cases the viewer eventually receives feedback"—how often, under what circumstances, exactly when? We need to know how close the "decoys" were to the true target. Who are the judges? This information is crucial because different judges with different backgrounds may classify differently.

The experiment is not duplicable. Even if you watched the experiment being done, you couldn't duplicate it for it's not clear what the author considers important and what not important in the set-up.

Even if it were possible to duplicate the experiment and get the same results, it's not clear that by chance alone the actual target would not receive each of the five ranks with equal likelihood. Perhaps this experiment would show the opposite. That is, the observational claims would not support the causal claim for which the author intends them to be used.

Example 7   The growth of living nerve cells in vitro
The immediate object of the following experiments was to obtain a method by which the end of a growing nerve could be brought under direct observation while alive, in order that a correct conception might be had regarding what takes place as the fibre extends during embryonic development from the nerve center out to the periphery.

The method employed was to isolate pieces of embryonic tissue known to give rise to nerve fibres, as for example, the whole or fragments of the medullary tube or ectoderm from the branchial region, and to observe their further development. The pieces were taken from frog embryos about three mm. long, at which stage, i.e. shortly after the closure of the medullary folds, there is no visible differentiation of the nerve elements. After carefully dissecting it out the piece of tissue is removed by a fine pipette to a cover slip upon which is a drop of lymph freshly drawn from one of the lymph sacs of an adult frog. The lymph clots very quickly, holding the tissue in a fixed position. The cover slip is then inverted over a hollow slide and the rim sealed with paraffin. When reasonable aseptic precautions are taken, tissues will live under these conditions for a week and in some cases specimens have been kept alive for nearly four weeks. Such specimens may be readily observed from day to day under highly magnifying powers.

> Ross Harrison, *Proceedings of the Society for Experimental and Medicine Biology*, vol. 4, 1907 (as quoted in *The Origins and Growth of Biology*, ed. Arthur Rook, pp. 159–160).

Analysis   This is the first method ever recorded for maintaining living cells outside the body. It is very much like the recipe from the *Dog & Duck*. Even for an expert it would have been difficult to duplicate it just from reading this.

62   Reasoning in Science and Mathematics

The morals of these examples

- It's very hard to describe an experiment clearly enough to duplicate it.
- What is described in an experiment is what needs to be duplicated. What is not described is deemed irrelevant to obtaining similar observations.
- What counts as duplicable is going to be relative to the particular scientific discipline. Expert knowledge in the area may make some descriptions very clear.
- What counts as close enough agreement for observations to be deemed replicable is going to depend on the particular scientific discipline.
- New experimental designs are often sketchily described, but they are accepted anyway because people go to the lab, see how it is done, then go back to their labs and do the experiment, and then pass that on to other people.

Experiments as the basis of correlations for causal reasoning

Experiments are often used to establish correlations from which cause and effect claims can be deduced. Statistical tests are used to show that a correlation which the experimenter has found is significant and not just a random occurrence.

Example 8   An experimenter flips a coin 4,318 times. It comes up heads 52% of the time. Statistical tests show that this is close enough to the expected outcome of 50% to be statistically insignificant.

Example 9   An experimenter flips a coin 4,318 times. It comes up heads 82% of the time. The statistical tests indicate that this is significant: the likelihood of it happening by chance is extremely small. And that it came up heads 47 times in a row once and 402 times in a row another time in the sequence is even less likely to be by chance.

So the experimenter comes up with some hypotheses: the coin is unbalanced; the person who flipped the coin has an ability to make the coin land on one side rather than the other; the room and atmosphere where it was done affects how it lands. These are causal explanations.

Yet despite the very great unlikelihood of these results happening, it still might be they occurred by chance. The expected probability of heads coming up 50% of the time is for the long run: if the experiment had been continued for another 10,972 times, perhaps the experimenter would have found that the percentage of heads came out much closer to 50%.

So she and others duplicate the experiment. One hypothesis requires that the person flipping the coin be the same. But what has to be duplicated is unclear: Does he have to wear the same shirt and trousers? Does he have to have bathed at the same time before the flipping? Another hypotheses requires that the room and situation in which the flipping is done be the same, though exactly what in the situation has to be duplicated is again unclear: Is the time of day important? Is the day of the year important?

If it turns out that the observations cannot be replicated, then there are two possibilities. Something quite unusual in the experimental situation might have been overlooked in the duplications. It really was a causal correlation, but neither the experimenter nor anyone else can figure out what condition is needed for that effect. Or the result was just chance, a random occurrence. The latter would generally be the accepted explanation for what happened, though if there is enough interest in the experiment others might try to vary the conditions of the duplication to try to replicate the observations.

The expectation of the scientific community is that if a description of an experiment is published which has importance for some area of science, then others will try to duplicate it and find whether the observations are replicable. This, it is believed, will winnow out the chance correlations from those that are significant for causal claims.

Example 10 When I saw a demonstration of what is known as the Mules operation for the prevention of blowfly attack in sheep, I realised its significance and my imagination was fired by the great potentialities of Mules' discovery. I put up an experiment involving thousands of sheep and, without waiting for the results, persuaded colleagues working on the blowfly problem to carry out experiments elsewhere. When about a year later, the results became available, the sheep in my trial showed no benefit from the operation. The other trials, and all subsequent ones, showed that the operation conferred a very valuable degree of protection and no satisfactory explanation could be found for the failure of my experiment. It was fortunate that I had enough confidence in my judgment to prevail upon my colleagues in other parts of the country, for if I had been more cautious and awaited my results they would probably have retarded the adoption of the operation for many years. p. 24

W.I. B. Beveridge, *The Art of Scientific Investigation*

Analysis One experiment, either to display a correlation or to disprove a conjectured one, is never enough to rule out chance, no matter the statistical tests it passes nor the skill of the experimenter.

Example 11 It is not at all rare for scientists in different parts of the world to obtain contradictory results with similar biological material. Sometimes these can be traced to unsuspected factors, for instance, a great difference in the reaction of guinea pigs to diphtheria toxin was traced to a difference in the diets of the animals. In other instances it has not been possible to discover the cause of the disagreement despite a thorough investigation. In Dr. Monroe Eaton's laboratory in the United States influenza virus can be made to spread from one mouse to another, but in Dr. C. H. Andrewes' laboratory in England this cannot be brought about, even though the strains of mice and virus, the same cages and an exactly similar technique is used. p. 24

W.I. B. Beveridge, *The Art of Scientific Investigation*

Analysis If only one duplication fails to replicate observations, then it might be chance. But if, as in this example, some replicate the observations and others don't, then researchers will try to find other causal factors that could explain the difference.

Lately, however, a disturbing pattern has emerged in attempts to replicate scientific findings. An experiment is done and a highly significant correlation is found. But as more and more duplications of the experiment are made, the correlation begins to decline until finally little or no statistically significant correlation is found in any of them.

One explanation of the *decline effect* is that it is simply regression to the mean. That is, the correlation in the original experiment occurred by chance, and with more duplications the measurements approach the actual probability, eventually showing little or no correlation. Another explanation is suggested by Jonah Lehrer.

Example 12 [Michael] Jennions, similarly, argues that the decline effect is largely a product of publication bias, or the tendency of scientists and scientific journals to prefer positive data over null results, which is what happens when no effect is found. The bias was first identified by the statistician Theodore Sterling, in 1959, after he noticed that ninety-seven per cent of all published psychological studies with statistically significant data found the effect they were looking for. A "significant" result is defined as any data point that would be produced by chance less than five per cent of the time. This ubiquitous test was invented in 1922 by the English mathematician Ronald Fisher, who picked five per cent as the boundary line, somewhat arbitrarily, because it made pencil and slide-rule calculations easier. Sterling saw that if ninety-seven per cent of psychology studies were proving their hypotheses, either psychologists were extraordinarily lucky or they published only outcomes of successful experiments. Jonah Lehrer, "The truth wears off"

Analysis But as Lehrer points out in this article, publication bias and regression to the mean together still don't explain why the decline occurs slowly over time, which several noted researchers have found in their own work.

Still, bias by researchers, even unnoted by them, can be a serious problem in interpreting data.

Example 13 One of the classic examples of selective reporting concerns the testing of acupuncture in different countries. While acupuncture is widely accepted as a medical treatment in various Asian countries, its use is much more contested in the West. These cultural differences have profoundly influenced the results of clinical trials. Between 1966 and 1995, there were forty-seven studies of acupuncture in China, Taiwan, and Japan, and every single trial concluded that acupuncture was an effective treatment. During the same period, there were ninety-four clinical trials of acupuncture in the United States, Sweden, and the U.K., and only fifty-six per cent of these studies found any therapeutic benefits. As [Richard] Palmer notes, this wide discrepancy suggests that scientists find ways to confirm their preferred hypothesis, disregarding what they don't want to see. Our beliefs are a form of blindness.

Jonah Lehrer, "The truth wears off"

Analysis Lehrer and Palmer do not consider the possibility that the practitioners of acupuncture in Asian countries, where it originated, are more highly skilled than those in the West, or that in the West only people more desperate for a cure because of a serious illness allow acupuncture to be done on them, or that there is a signficant placebo effect in Asian countries.

The morals of these examples
- Never trust the first or even the first few experiments that claim to establish a significant correlation.
- To reason well, imagine the possibilities.

Mathematics as the Art of Abstraction

Mathematics, like the sciences, proceeds by a process of abstraction, so that mathematical claims like scientific claims are neither true nor false, but only true or false in an application of the theory to which they belong. A proof in mathematics is meant to show that a claim follows from the assumptions of a particular mathematical theory.

Introduction 68
Analogies and abstraction in science 68
Mathematics as the art of abstraction
    Counting and addition 69
    Subtracting 71
    Irrationals 72
    Complex numbers 72
    Plane geometry 72
    Group theory 73
    Ring theory 74
    Transfinite ordinals 76
Mathematical proof 76
Mathematical proofs are arguments 78
Comparing two proofs of a simple claim in arithmetic . . . 81
A proof of Pythagoras' theorem, and progress in mathematics . . 83
Formal proofs 85
Mathematics as pure intuition 85
Set theory and the existence of infinities 86
Mathematical proofs as explanations 88
The utility of this story 93
Grounding our stories of mathematics 93
Appendix 1 Mathematics as an innate ability 95
Appendix 2 Mathematics as used in science 97
Appendix 3 Comparisons to other views of mathematics
    Structuralism 98
    Deductivism 99
    Mathematics as an empirical science 103
Notes 105

Introduction

I'd like to tell you a story, a story of how I understand mathematics.

There are so many stories already: the platonist's, the formalist's, the constructivist's, the structuralist's, ... But each of those fails to answer satisfactorily at least one of the following questions:

- How do we create and know mathematics?
- How does mathematics compare to our other intellectual activities, particularly science?
- What is mathematical intuition?
- What is a proof in mathematics?
- Is a good proof in mathematics also a good explanation?
- What is mathematical truth, and are mathematical truths necessary?
- How is it that mathematics is useful in our daily lives and in science?

Any story of mathematics should answer all of these. But a good story should also:

- Be consistent with how we actually do mathematics.
- Be useful to mathematicians, leading to new and interesting work in mathematics.

I hope my story is a good one.

Analogies and abstraction in science [1]

One of the fundamental ways of reasoning about what passes in our lives is reasoning by analogy.

> A comparison becomes *reasoning by analogy* when a claim is being argued for: on one side of the comparison we draw a conclusion, so on the other side we can draw a similar one.

This situation or thing is just like that; since we can draw this conclusion about the first, we are justified in drawing a similar conclusion about the second.

Such reasoning is not good until we can say in each particular case what we mean by "is just like" or "similar." No two things, no two situations, no two experiences are exactly the same. We pay attention to some similarities and ignore the differences. If the differences don't matter, or rather, if they don't matter too much, then we are justified in drawing similar conclusions. The point of an analogy is to force us to be explicit about that justification, setting out, if we can, some general claim under which the two sides fall and from which the conclusions follow. Such reasoning is pervasive in our daily lives.

Science sometimes proceeds by analogy. But scientists always proceed by abstracting: choosing some aspect(s) of experience to pay attention to and claiming, perhaps implicitly, that all other aspects of experience in these kinds of situations don't matter. What we pay attention to gives us the constraints for saying whether a claim is true or false.

A scientific theory is true in the context of what we pay attention to but is false in that it does not take into account all of the world. The hypotheses of scientific theories act as conditions for where the theory can be applied. When we "falsify" a scientific theory, we do not show that it is false. What we show is that it is not applicable to the experience described in the falsifying experiment. Then we try to describe carefully what kind of experience that may be, adding more conditions to our theory in the form of further premises. Thus, we can continue to use Newton's theory of motion even though, it is said, it is false; we use it so long as the objects we are investigating are not too large nor too small and are not going too fast; we use it where it is applicable.

The abstractions that comprise science are not false, nor are they true. They are schematic claims until we say what we are paying attention to.

Mathematics as the art of abstraction

Mathematics abstracts from experience, too, only much more than any science.

Counting and addition
We first have numbers as adjectives: one dog, two cats, eighteen drops of water, fourteen sonatas, forty-seven ideas about mathematics. Numbers are labels.

With practice, repeating and learning the rules for naming and writing new numerals, the counting numbers become a measuring stick we carry in our head. We count off objects to find how many, as we use a tape to measure off lengths of objects. There is a definite length we cannot go beyond in measuring with our tape, but there is no definite limit we perceive in how far we can go with counting.

Counting to find how many, however, is not learned independently of using counting for addition. Consider how we learn $3 + 5 = 8$. We take something like pebbles, or pieces of candy, or dots on a paper and count:

    • • •   • • • • •
    1 2 3   1 2 3 4 5
    1 2 3   4 5 6 7 8

We show that when we put the two sequences of counting together into one sequence we get the result that the last item in the new sequence is assigned "8". Then, after a lot of practice, it seems to us that the same results of counting would apply to any other objects like these; the two ways of counting aren't idiosyncratic to just these dots or pieces of candy. We learn to ignore what doesn't matter—we abstract—and get a theory of addition.

We've been doing this so long, we learn it so early, it is so much a part of our culture that we don't see this as a model. Surely "$1 + 2 = 3$" is true.

But 1 drop of water + 2 drops of water ≠ 3 drops of water. "$1 + 2 = 3$" is not a truth about the world; it is one of the claims that is needed to apply in a situation in order for arithmetic to be applicable there. We can't use arithmetic for drops of water when we put those together.

Our abstraction from counting is applicable or not. Our "truths" of arithmetic say what follows from our abstraction and hence when our model is applicable, just as in any reasoning by analogy or abstraction. Arithmetic is an application of measuring, and we must measure correctly. We have to learn (as children) how to apply the model.

It seems to us, once we get the hang of it, that counting-addition is univocal. But that only says we have made it very clear how to apply it. There is no reason to think ahead of time that we will never encounter another situation in which it seems by all we have done so far that our counting-addition is wrong. If we do, we will surely restrict counting-addition not to apply to such a case, saying that what we thought were objects are not things, for only things can be counted.[2] That is, "$1 + 2 = 3$" cannot be falsified not because it is a necessary truth but because we preserve it to be true by applying it only in cases in which it is true.

Then we take numbers as nouns. We reify our abstracting: the end of the process of abstraction—paying attention to only some of our experience—begins to be treated as a thing, an abstract thing. We abstract from counting and addition to get the natural numbers: 1, 2, 3, ...[3]

To say that "1 + 2 = 3" is true is to say that numbers—not as adjectives or as arising from counting, or as abstracted from those uses—are actual things about which we reason. That is not incompatible with what I've said, but it gives us no insight into how we create and do mathematics and why mathematics is applicable.

Subtracting

When we learn how to count and add we also learn how to subtract. We abstract from the process of taking away objects ("Hey, when you took six pieces of candy that left me with only five!") to get a theory of subtraction. It may sound odd to call addition and subtraction "theories" when they are just part of what we do every day. But they are theories just as much as Newton's laws of motion in that they abstract from our experience.

When we do subtraction along with addition, we find that our calculations—that is, the working out of claims without reference to the things to which they might apply—go a lot more clearly and smoothly if we have some "things" called zero and negative numbers. That's how and why those were first introduced. They flesh out our abstractions. They make the calculations easier. They don't seem to apply to anything.

And hence we feel uneasy about them. If they aren't abstractions from our experience, how can we trust that the calculations in which we use them give results that are applicable? When negative numbers were first introduced in the 16th century, they were suspect. How can we understand the equality of ratios? How can 1: –5 = – 4: 20 when in the first ratio 1 is "larger" than –5, while in the second – 4 is "smaller" than 20? Objections and questions about the legitimacy of their use continued until the 19th century, when they were given a visual/physical interpretation.[4]

As in any scientific theory, if we introduce new "entities" that do not arise by abstraction, there will be, and should be, objections about their use in the theory. When we can see a path of abstraction to them, we begin to have confidence in our theory.

Irrationals

Irrationals were not introduced like negative integers. Irrationals were always part of the abstraction of space we call geometry. Or rather, points on a line were always part of that abstraction, and some of those, it was discovered, couldn't be measured by ratios of integers. The measuring of those points then became viewed as things: irrational numbers.

Complex numbers

The square root of −1 was introduced into algebraic calculations because it facilitated calculations and gave new results that could be checked by older methods. Such calculations were challenged by many mathematicians as being fantasy, as having no physical counterpart, as not reliable. Yet mathematicians continued to make those calculations in their work because no contradictions arose when they were used correctly, that is, according to the rules that were eventually determined for them.

Eventually complex numbers became accepted because they were given a visual/physical representation as points on a plane. That clear path of abstraction made us feel confident that their use was legitimate.

Plane geometry

Euclidean plane geometry speaks of points and lines: a point is location without dimension; a line is extension without breadth. No such objects exist in our experience. But Euclidean geometry is remarkably useful in measuring and calculating distances and positions in our daily lives.

Points are abstractions of very small dots made by a pencil or other implement. Lines are abstractions of physical lines, either drawn or sighted. So long as the differences don't matter—that is, so long as the size of the points and the lines are very small relative to what is being measured or plotted—whatever conclusions that we draw will be true. Defining a line as extension without breadth is an instruction to use the theory only when we can ignore the breadth of the line.

No one asks (anymore) whether the axioms of Euclidean geometry are true. Rather, when the differences don't matter, we can calculate and predict using Euclidean geometry. When the differences do matter, as in calculating paths of airplanes circling the globe, Euclidean plane geometry does not apply, and another model, geometry for spherical surfaces, is invoked.

Euclidean geometry is a mathematical theory, which, taken as mathematics, would appear to have no application since the objects of which it speaks do not exist in our experience. But taken as a model it has applications in the usual sense, arguing by analogy where the differences don't matter.

Group theory
Consider a square.

If we rotate it any multiple of 90° in either direction it lands exactly on the place where it was before. If we flip it over its horizontal or vertical or diagonal axis, it ends up where it was before. Any of these operations followed by another leaves the square in the same place, and so is the same result as one of the original operations.

To better visualize this, imagine the square to have labels at the vertices and track which vertex goes to which vertex in these operations. There is one operation that leaves the square exactly as it is: do nothing. For each of these operations, there is one that undoes it. For example, the diagonal flip that takes vertex **c** to vertex **a** is undone by doing that same flip again; rotating the square 90° takes vertex **a** to vertex **b**, and then rotating 270° in the same direction returns the square to the original configuration.

Already we have a substantial abstraction. No rotation or flip leaves the square drawn above in the same place because we can't draw a square with such exact precision (nor can a machine), and were we (or a machine) to cut it out and move it, it wouldn't be exactly where it was before. We are imagining that the square is so perfectly drawn, abstracting from what we have in hand, choosing to ignore the imperfections in the drawing and the movement. If we like, we can then talk about abstract, "perfect" objects, platonic squares, of which the picture above is only a suggestion, but that seems to be only a reification of our process of abstracting.

Now consider several objects lined up in a row. We permute, that is exchange, places of some of them, again leaving them lined up in a row. Any permutation followed by another is a permutation, a way to get the objects lined up in a row again. There is one permutation that leaves everything unchanged: do nothing. For any permutation we can undo it by reversing the replacements.

Consider, too, the integers and addition. There is exactly one integer which, when we add it to anything leaves that thing unchanged, namely, 0. And given any integer there is another which when added to it yields back 0, namely the negative of that integer.

Mathematicians noted similarities among these and many more examples. Abstracting from them, in the sense of paying attention to only some aspects of the examples and ignoring all others, they arrived at the definition of "group." A typical definition is:

A *group* is a non-empty set G with a binary operation • such that:

i. For all a, b in G, a • b is in G.

ii. There is an e in G, called the *identity*, such that for every a in G, a • e = e • a = a.

iii. For every a in G, there is an a^{-1} in G, called the *inverse* of a, such that $a • a^{-1} = a^{-1} • a = e$.

iv. The operation is *associative*, that is, for every a, b, c in G, (a • b) • c = a • (b • c).

We can prove claims about groups, such as that there can be only one identity in a group. Such a proof does not show that "There is only one identity in a group" is true — not, that is, unless we reify our abstraction into abstract things called groups. But such a proof does give us a true claim:

"There is only one identity" follows from the assumptions of group theory.

If the assumptions of group theory apply to some thing/situation/part of our experience/process, then the claim about the identity of the group must apply, too. If that claim fails to apply, then we know that group theory is not applicable in that case, not that group theory is false.

Ring theory

We have the integers with addition and multiplication. We have the real numbers with addition and multiplication. We have the rationals with addition and multiplication. These and many more examples led to the notion of a ring in mathematics. We can abstract from our abstractions.

But don't we mean by "abstract" to ignore and pay attention to certain aspects of our experience?

The practice of doing mathematics is part of our experience, too. Mathematical abstractions are part of our intersubjective mental life, like laws and sonatas. Our intersubjective mental activities are also suitable subjects for abstraction. Mathematics is a human activity.

We can abstract from our abstractions, going further and further in ignoring aspects of the ordinary experience of our daily lives. Some of us have great pleasure in considering and reasoning in such highly abstract subjects. But the pleasure is merely aesthetic until applications of the abstractions are found, relating abstract subjects such as algebra and plane geometry as René Descartes did.

Perhaps you've heard the phrase "the unreasonable effectiveness of mathematics" used as shorthand for saying that it is very odd that highly abstract theories of mathematics developed solely within the context of other abstractions can so often be applicable.[5] Such a view is mistaken. As mathematicians know, it is rare for a mathematical theory of abstractions of abstractions of abstractions to be applicable to our daily experience, as opposed to just the experience of doing mathematics of a few mathematicians. And when such applications can be found, why should it be so shocking? Yes, the theory was derived from abstracting from abstracting, perhaps many levels. An application is the result of going back along that path of abstraction. And that can only be if what the theory pays attention to and what it ignores is aptly chosen. We can make a theory that ignores almost everything and pays attention to bizarre or abstruse or little-used parts of our experience. Sometimes, by chance, those aspects that are considered turn out to be important in other parts of our experience. In those few cases, the "pure" mathematics is applicable to more of our ordinary lives. We have hit upon a good analogy.[6]

The great mathematicians—those who have some insight into what claims follow in some mathematical theory or who create a theory joining parts of mathematics never before considered similar, who make abstractions of abstractions well—are said to have great intuition, mathematical intuition. That intuition is no different from the intuition that leads a wise person to draw an analogy between dogs and humans in arguing for the humane treatment of animals, or the intuition of the wise person who first "sees" that light can be understood as waves. In my own experience I find that the intuition I had in seeing the general outlines of this paper before I began writing it, and the intuition I had in proving a new theorem about degrees of unsolvability, and the intuition I had in writing one of my plays are the same mental activity, differentiated only by the subject matter. I see a general picture, I see a few of the details, I begin with that "vision" or "insight" ahead of me, and I fill in the details, often arriving at something quite different from what I first imagined.

We do not understand how such intuition works. But there is no reason to think that mathematical intuition is something different in kind rather than different only in its subject matter.

Transfinite ordinals
We have our theory of counting with the natural numbers. We can count forever and never reach the end.

But we can imagine that there is an end, even if we are not "able" to reach it, just as there is an end to every counting we do in our lives. By analogy we postulate an end to the sequence of natural numbers and call it ω (omega). Then we can continue our counting: $\omega + 1, \omega + 2, \omega + 3, \ldots$. And since we've done it once, we'll do it again: assume an end to that counting, $\omega + \omega$. And then we can continue such counting forever.

This sounds like pure fantasy. What's to tell us that this will "work" in the sense of never leading to contradictions? And why bother? What's the use?

We can describe such ways of counting in a constructive manner as arrays of natural numbers: one sequence is ω; two sequences, with every number in the first coming before every number in the second is $\omega + \omega$; .... When we see a picture of this, we can see a path of abstraction. Moreover, such ways of counting can be seen to correspond to more and more complicated forms of proof by induction.[7]

But when someone considers extending the counting beyond what we can constructively describe, indeed into an infinite that is beyond what could be "counted" in any sense, we have more serious doubts. Why is this acceptable? Why is this theory based on analogy and postulating new kinds of doings even consistent?

Mathematical proof
As I mentioned in the discussion of group theory and will make clearer now:

*A mathematical theorem does not show that a claim is true.
It shows that the claim follows from the assumptions of the theory.*

When we give a proof in Euclidean plane geometry of the claim that there cannot be two distinct lines through a point parallel to another line not through that point, we are not showing that claim is true. We've seen that it doesn't even make sense to say that it is true but only true of something, in the sense of applicable. Our proof is good if it is a good argument to establish that the claim about parallel lines follows from the axioms of Euclidean plane geometry.[8]

Mathematicians do not say, "There cannot be two distinct lines through a point parallel to another line not through that point." They say, "In Euclidean geometry, there cannot be two distinct lines through a point parallel to another line not through that point." Mathematicians do not say "There is a unique identity." They say, "In any group the identity is unique," invoking, establishing thereby that the claim is "true" in the theory of groups.

*The truths of mathematics are truths about inferences.
Mathematics is about what follows from what in our abstractions.*

Yet many mathematicians and philosophers say not only that mathematical claims are true but that they are necessarily true: there is no possible way such a claim could be false.

There is indeed a necessity in our mathematics, but it is not the necessity of a claim such as "$2 + 2 = 4$." The necessity is that the claim must follow from the assumptions of the theory. There is no way that the axioms of Euclidean geometry could be true and the claim about parallel lines false. There is no way that the assumptions of group theory could be true and the claim that the identity of a group is unique be false. We demand of a mathematical proof that it establish that the inference from the assumptions of the theory to the claim to be proved is a *valid inference*: it is impossible for the premises to be true and the conclusion false (at the same time and in the same way).

This requirement on mathematical proofs goes back to before Euclid. It does not rely on any analysis of the forms of claims. It invokes only the notions of possibility and truth. What the notion of validity requires, and what it guarantees when it holds of an inference, is that if the mathematical claims that are the premises are true in any application, then the conclusion will be true in that application, too. The following picture summarizes this schematically:

A Mathematical Proof

Assumptions about how to reason and communicate.

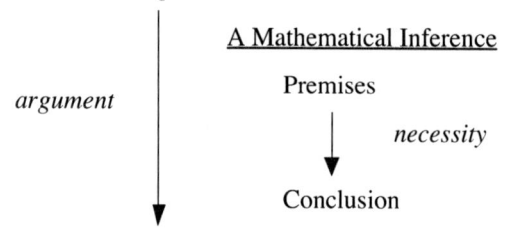

```
                        A Mathematical Inference
                           Premises
  argument                    |
                               |    necessity
                               ▼
            |              Conclusion
            ▼
```

The mathematical inference is valid.

A mathematical inference is a statement of a theorem, for example:

The axioms of Euclidean plane geometry.
Therefore the parallel lines postulate.

The conclusion of a mathematical proof is that the mathematical inference is valid, though mathematicians are rarely so careful as to say that explicitly. Rather, they just show the inference is valid. The mathematical proof as a whole must be a good argument for that.

Mathematical proofs are arguments

Reuben Hersh, a mathematician reflecting on his and others' work in *What is Mathematics Really?* characterizes mathematical proofs [9]:

> Mathematical discovery rests on a validation called "proof," the analogue of experiment in physical science. A proof is a conclusive argument that a proposed result follows from accepted theory. "Follows" means the argument convinces qualified, skeptical mathematicians. Here I am giving an overtly social definition of "proof." p. 6

Hersh, as many others, has conflated the two arrows in the picture above. The mathematical proof must be a good argument, and that can be loosely described as one that convinces qualified, skeptical mathematicians. But the "follows from" that must be established is that the mathematical inference is valid.

We can say a great deal about what constitutes a good argument. To begin:

An *argument* is an attempt to convince someone, possibly yourself, that a particular claim, called the *conclusion* is true. The rest of the argument is a collection of claims, the *premises*, which are given as the reason for believing the conclusion is true.

The following are necessary conditions for an argument to be good:

> The premises are plausible.
> The premises are more plausible than the conclusion.
> The conclusion follows from the premises.

Plausibility is a measure of how much reason we have to believe that a claim is true. In mathematical proofs it is not the assumptions of the mathematical inference that must be plausible. After all, how can we say that the axioms of Euclidean plane geometry are plausible when they aren't even true or false? Some of them aren't even simpler than the claim about parallel lines.[10]

The plausibility conditions apply to the mathematical proof as a whole: the assumptions about the nature of reasoning and abstractions and how we communicate must be plausible and more plausible than the conclusion that the mathematical claim follows from the assumptions of the theory.

Many times in the history of mathematics the assumptions of mathematical proofs—the part labeled "argument"—have been questioned. At one time it seemed (or still seems) very dubious that we could reason using negative numbers, or the square root of -1, or that we could invoke infinities, or use the method of proof by contradiction, or use the law of double negation, or invoke the axiom of choice. In those cases, though no one doubted that the mathematical inference was indeed valid if given those assumptions, they doubted that the proof, the argument using those assumptions, really established that the inference was valid. They found the premises of the argument—not of the inference—dubious.

The premises of mathematical proofs, the claims about reasoning and communication, the assumptions about mathematics that we use in our everyday work in mathematics are usually left unsaid; they are treated briefly in undergraduate texts and even more briefly in undergraduate mathematics courses. They become of interest to mathematicians only when paradoxes or wrong proofs arise in some area of mathematics. Uncovering and making those assumptions explicit so that we can debate them is what logicians and philosophers of mathematics do. Logicians and philosophers then join in the debates with working mathematicians, for it is the working mathematicians who will finally decide if some method or assumption is acceptable in the canon of mathematics.

Two examples about what assumptions are currently acceptable in mathematical proofs illustrate these points.

Some mathematicians have given proofs that certain very large numbers are prime by setting out probabilistic analyses. They show that within a very small possibility of being wrong a particular number is prime. But they then want to say more: this suffices to show that the number is prime. After all, they say, if it is certainty we want from our proofs, many mathematical proofs that are very long and have many steps left to the reader are less certain than the small probability of error of their straightforward probabilistic proofs.

But it is not certainty that is at issue. The issue is whether the inference in a mathematical proof must be shown to be valid, not whether the argument, the mathematical proof, is convincing. If we accept probabilistic inferences in mathematics, then mathematics is no different from any science, for all sciences use strong arguments and not just valid ones to show that a claim follows from the assumptions of their theories. At present there is little division in the mathematical community about this issue: almost everyone agrees that a mathematical proof must show that the inference is valid.

The other new method of proof utilizes computers to evaluate many complicated cases that could not be done by hand, eliminating each as a possible counterexample, and concluding a claim such as the four-color theorem. Mathematicians are hesitant to accept such proofs because they rely on our trusting that the computer software is right and that the computer itself is functioning correctly. How can that be part of mathematics? In response, it is again said that very long proofs that leave many steps to the reader and are accepted on the word of one or two referees are much more dubitable than such computer proofs.

Here the issue is not whether a mathematical proof should establish that an inference is valid, for the computer proof is claimed to do just that. The issue is about what counts as a good argument in mathematics.

Mathematical arguments, just like arguments in our daily lives, leave much unsaid. And of what is said, much is only hints or sketches, with lots explicitly left to the reader. We accept such arguments because they are a form of communication. We can see how to understand and evaluate an argument made by a person, filling in the gaps when needed. When we cannot fill in the gaps, when questions cannot be answered, we reject the argument as a mathematical proof.

In contrast, a proof by computer can only be followed step by step in the hopes that we can see how each step is used in the proof, and that is impossible when there are so many steps that the prover had to have recourse to a computer. We cannot imagine the intention of the computer as a guide for how to repair a proof, for computers do not have intentions.

We can, however, try to verify that the program run by the computer does what it is intended to do. We can perform tests on a few inputs where the outputs are already known, we can examine how the program is written, and declare that the program is correct. It is possible that the program might not be, but the chances of that, it is believed, are small enough considering the cost of more extensive checking. But that, still, leaves us only with accepting or rejecting a proof by a computer, not understanding it as one would a human communication. The debate on the acceptability of computer programs in mathematical proofs continues to divide the mathematical community.[11]

Comparing two proofs of a simple claim in arithmetic

<u>Proof 1</u> $1 + 2 + \cdots + n = 1/2 \, n \cdot (n+1)$

Proof $1 = 1/2 \cdot 1 \cdot (1+1)$. This is the basis of the induction.

Now suppose that $1 + 2 + \cdots + n = 1/2 \, n \cdot (n+1)$. This is the induction hypothesis. Then:

$$1 + 2 + \cdots + n + (n+1) = [1/2 \, n \cdot (n+1)] + (n+1).$$

So $1 + 2 + \cdots + n + (n+1) = 1/2 \, (n^2 + n) + 1/2 \, (2n+2),$

so $1 + 2 + \cdots + n + (n+1) = 1/2 \, (n^2 + 3n + 2),$

so $1 + 2 + \cdots + n + (n+1) = 1/2 \, (n+1) \cdot (n+2).$

That is, $1 + 2 + \cdots + n + (n+1) = 1/2 \, (n+1) \cdot ((n+1) + 1).$

Some proof like this is what all mathematicians have encountered in learning mathematics. We are told that anything less rigorous than this doesn't count as a proof. We are told it establishes that "$1 + 2 + \cdots + n = 1/2 \, n \cdot (n+1)$" is true.

But the proof does not establish that "$1 + 2 + \cdots + n = 1/2 \, n \cdot (n+1)$" is true. What it shows is that the claim follows if we accept the method of proof by induction. But isn't that true? True of what—

our abstraction? No. The method of proof by induction is a condition of our theory: it is part of what establishes the subject of arithmetic. That method is not obvious to students until they have been drilled on it; it is very hard to use. Any claim that follows from the method of proof by induction on the natural numbers must be applicable to any instantiation of the theory. If such a claim fails, then that's not what we are talking about.

This is what the ultra-constructivists who deny induction miss.[12] Their program is not about what the natural numbers are, nor the right way to reason about them, nor disproofs of induction. They are going back to counting and deciding not to make the abstraction from our abilities to say that we can "count" forever.[13] They propose a more "realistic" theory of counting and arithmetic, where a theory can be said to be more realistic than another if it abstracts less from the same experience(s).

<u>Proof 2</u> $1 + 2 + \cdots + n = 1/2\, n \cdot (n + 1)$

● ○ ○ ○ ○ ○ ○
● ● ○ ○ ○ ○ ○
● ● ● ○ ○ ○ ○
● ● ● ● ○ ○ ○
● ● ● ● ● ○ ○
● ● ● ● ● ● ○

Is this a proof? When I saw it I felt for the first time that I understood and could believe "$1 + 2 + \cdots + n = 1/2\, n \cdot (n + 1)$". Before, I only knew from memory that there is a proof using a manipulation of symbols that I could reconstruct fairly easily. Each time I did that symbolic proof I saw indeed that the claim followed by induction. But here I could see that it was true, really true.[14]

True? The picture convinced me, without any recourse to induction, that the equality will apply to any things to which our basic model of counting and arithmetic apply. I see from the picture that any things I can count up to some number, call it n, can be put, along with other such things, in an array that justifies the equality by counting again. And a 6 by 7 array was enough to convince me.[15]

But how do we know that we'll always get the same result, regardless of how large we expand the diagram? We know in the same way that we know "3 + 5 = 8" will be true for any selection of objects to which we wish to apply our methods of counting and addition. We can give a proof of "3 + 5 = 8" from axioms for arithmetic, but that is less convincing and assumes a great deal more as a theory of addition.

The first proof shows that the equality follows from the assumptions of our more general theory of counting and arithmetic in which we also accept proof by induction; the assumptions about reasoning, though not about the nature of our abstractions, are relatively explicit. This second proof by picture leads us to believe that the equality applies to anything to which our more basic theory of counting and arithmetic apply without invoking induction; the assumptions about the modeling are relatively clear, but the methods of proof are not. In neither case are we showing that a claim is true; we are showing that it follows, that it is part of our theory. The truths of mathematics are truths about inferences and applications of theories.

A proof of Pythagoras' theorem, and progress in mathematics

Consider another proof by diagram, this time of Pythagoras' theorem.

First, we can represent the product of two numbers **a** and **b** by a rectangle with sides of length **a** and **b**. Then we have the identity:

$$(a + b)^2 = a^2 + b^2 + 2ab$$

Now we note the area of a right triangle with sides **a** and **b** is $1/2$ **ab** :

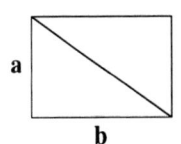

Then we can get Pythagoras' theorem: the square of the length of the hypotenuse in a right triangle is the sum of the squares of the lengths of the legs. All we need do is calculate the area of the large square below by the two different methods:

$a^2 + b^2 = c^2$

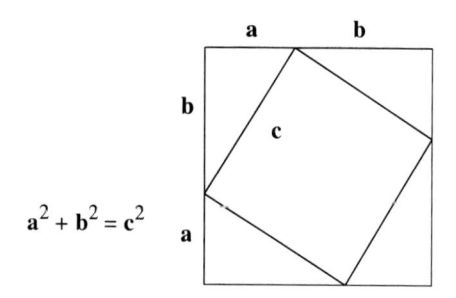

It is clear to me once I have understood the diagrams that it is impossible to come up with a triangle that would not satisfy the identity.[16]

Yet some say that these diagrams do not constitute a proof. Rather, it is only in Euclid that we find a real proof of Pythagoras' theorem. That view, I believe, is a reflection of the desire to make our implicit assumptions explicit. Geometric assumptions necessary for proving Pythagoras' theorem are explicitly set out in Euclid. Or at least they seemed to be for over two millennia. But now the same criticism of Euclid can be given as is given of the diagrams: too much was left as unstated assumptions. The first time we had a real proof of Pythagoras' theorem, it is said, was when Hilbert formalized Euclidean geometry.

What is a proof depends on what we tolerate as implicit assumptions. If progress in mathematics means (at least in part) making more explicit the assumptions by which we prove results, then it is not clear that Hilbert's system is progress over the diagrams because the assumptions he has made explicit seem to go far beyond what was in our use of the diagrams. What progress there is in mathematics depends more on the assumptions that are being made explicit having greater generality and applicability. That is, progress in mathematics in this sense depends on greater generality through abstraction.[17]

This is not, however, the exclusive kind of progress in mathematics. Solving a difficult and long-known problem counts as progress, too. But universally it is acknowledged that such a solution is not of much interest, that is, it is not really considered progress if it does not employ new methods that allow for the solution of other problems or a generalization of the original problem. That is what is frustrating about evaluating the use of computers in the proof of the four-color theorem: nothing new is used in the proof, no new generalities or ideas, just the brute force of the computer working through many, many cases.

Formal proofs

Some mathematicians think an objective standard can be given for what counts as a proof in mathematics. They say an argument counts as a mathematical proof only if it could be formalized, by which they mean within a system of formal logic.[18]

In *Classical Mathematical Logic* I provide a derivation from the axioms of group theory in first-order logic of the claim "The identity is unique." But that is not a mathematical proof. I give the mathematical proof first and then argue that the formalization is apt.

A proof in a fully formal system of logic that a claim follows from some axioms is not a proof in mathematics. It is evidence that can be used in a mathematical proof: Why should I believe that this claim follows from these others? Because—and we point to the formal proof. It is the pointing that is crucial. It relies on many assumptions, most particularly that the formal system chosen for the formalization is an apt model of reasoning and a good one for this mathematical proof, and that the steps that have been added—for there are always steps that have to be filled in—are appropriate.[19] We cannot fully formalize the argument that constitutes the mathematical proof without leading to an infinite regress.

We cannot replace proofs in mathematics with formal proofs, though we can use formal proofs as evidence in mathematical proofs. Formalizing mathematical proofs can lead to uncovering or clarifying assumptions behind such informal proofs and seeing how or whether such assumptions are needed.

Mathematics as pure intuition

At the other extreme, some say that mathematics is entirely subjective. The noted mathematician R. L. Wilder says:

> What is the role of proof? It seems to be only a testing process that we apply to these suggestions of our intuition. . . .
> Obviously, we don't possess, and probably will never possess, any standard of proof that is independent of time, the thing to be proved, or the person or school of thought using it.
> "The Nature of Mathematical Proof," pp. 318–319

The celebrated mathematician G. H. Hardy seems to concur:

> There is, strictly, no such thing as mathematical proof; we can, in the last analysis, do nothing but point; . . . proofs are what Littlewood and I

call *gas*, rhetorical flourishes designed to affect psychology, pictures on the board in the lecture, devices to stimulate the imagination of pupils. This is plainly not the whole truth, but there is a good deal in it.
"Mathematical Proof," p. 18

And the mathematician and philosopher L. E. J. Brouwer says,

> In the construction of [all mathematical sets of units which are entitled to that name] neither the ordinary language nor any symbolic language can have any other rôle than that of serving as a nonmathematical auxiliary, to assist the mathematical memory or to enable different individuals to build up the same set.
> "Intuitionism and Formalism," p. 81

To do mathematics we must use our intuition. But that does not mean that mathematics is subjective. It is intersubjective, like law, like drama, like etiquette. To appeal to each individual mathematician's intuition leaves us no criteria at all for judging whether we have a proof, just as appealing to only a judge's intuition gives no standard for what is legal. And we know at least one clear standard for mathematical proofs: they must establish that an inference is valid.

The view of mathematics as subjective introspection leaves it difficult for us to explain how we learn mathematics, how we judge whether what we have is a proof, what mathematical truth is, ... All the questions with which we began this paper remain unanswered. Though intuitionists who followed Brouwer have created a full and rich mathematical theory, they did so by denying his quote above: they use and work with proofs just as all other mathematicians do.

Set theory and the existence of infinities

Given any few small objects, we can collect them together. Given any things we can describe, we can make up a description of all of them at once. So proceeding by analogy, mathematicians in the 19th century assumed that we can "collect" any things whatsoever into a new entity called the "set" of those things.

We can describe a beautiful theory that way that has many applications. But it leads to contradictions, such as the set of all sets that are not elements of themselves. Not every beautiful way of postulating new things in analogy with old ones is good.

But the utility of such a theory is so desirable that mathematicians worked to rescue it, modifying the analogy to say that only certain ways of collecting are acceptable. And thus we have modern set theory. And we hope that it is consistent, for it is a very useful high-level abstraction in which we can codify and compare many areas of mathematics.

The assumptions of our new set theory countenance "collecting" all natural numbers into one set. We can also "collect" all points on the line into one set—as if a line that is finite but forever extendible were actually extended and completed as an infinite thing.

That makes a lot of mathematicians uneasy, from the ancients, through the 17th century, and continuing today. Intuitionists and constructivists deny that such analogies are legitimate in mathematics. They can see no path of abstraction that leads to such a fantastical analogy. Those who use set theory say it should be accepted because it is fruitful and (so far, it seems) consistent.

The utility and consistency of a mathematical theory, some argue, are sufficient for us to investigate it. Indeed, not even utility but just a sense that the theory is beautiful has been enough in the past century for mathematicians to publish papers on new theories.

But, as we've learned from Kurt Gödel, it is rare that we can prove a theory to be consistent.[20] So we try to relate it to other theories we know arise from a path of abstraction and for which we have inductive evidence of consistency: no contradiction has arisen in the many years that many mathematicians have worked in that theory.

Some go farther. They say that consistency of a mathematical theory is all that is needed for us to conclude that the things of which it speaks exist.

> In particular, in introducing new numbers, mathematics is only obliged to give definitions of them, by which such a definiteness and, circumstances permitting, such a relation to the older numbers are conferred upon them that in given cases they can definitely be distinguished from one another. As soon as a number satisfies all these conditions, it can and must be regarded as existent and real in mathematics. Georg Cantor.
> *Grundlagen einer allgemeinen Mannigfaltigkeitslehre*, p. 182

> A mathematical entity exists, provided its definition implies no contradiction, either in itself or with the propositions already admitted.
> Henri Poincaré, *The Foundations of Science*, p. 61

> If the arbitrarily given axioms do not contradict one another with all their consequences, then they are true and the things defined by the axioms exist. This is for me the criterion of truth and existence.
> David Hilbert, letter to Frege of December 29, 1899 [21]

The infinities beyond infinities of set theory exist, they say. And they often say so without the qualifier "if our set theory is consistent." They have gotten used to working in the theory and thinking of these objects, so how could they not exist? If the theory is contradictory, they feel that they can modify it once again to retain their world of abstract objects. They do not consider that if they modify their theory, the objects of which it speaks might not be the ones they had been thinking of earlier.

But we do not need that mathematical objects postulated by our theories exist in order for our theories to be used and to be useful. We only need that the theory is consistent, for then we can act *as if they exist*: it is not logically impossible for them to exist. And possibilities are all we need in order to reason about mathematical inferences in our mathematical proofs: an inference is valid if there is no *possible way* for the premises to be true and conclusion false.

Some find it remarkable that our theories are consistent and cite that as evidence that mathematical claims are indeed true or false.[22] But the process of abstraction, ignoring some of our experience and focusing on just part of it, is not likely to lead to an inconsistency. It is only when we postulate something in addition to our experience that we risk inconsistency. The great difficulty in analyzing processes of abstraction is distinguishing between those cases where we only ignore certain aspects of our experience, as with addition and multiplication of counting numbers, and those cases where we postulate some additional ability or capacity of ourselves or some extension of our experience, as in set theories that allow for infinite collections.[23]

Mathematical proofs as explanations

It is often said that a good mathematical proof does more than just show that a mathematical claim is true; it provides a good explanation of why the claim is true. We want to know not just that "$1 + 2 + \cdots + n = 1/2\, n \cdot (n + 1)$" is true, but why it is true.[24]

An explanation in this sense is characterized as follows:

> An *inferential* explanation is a collection of claims that can be understood as "E because of A, B, C, . . .". The claims A, B, C, . . . are called the *explanation* and E is the claim being explained. The explanation is meant to answer the question "Why is E true?"

For an inferential explanation to be good, the inference from A, B, C, . . to E must be valid or strong. And the claim being explained must be highly plausible: we do not explain anything we do not already believe.

To view a proof of a mathematical claim as an explanation raises the same problems as viewing a mathematical proof as an argument for that claim: we must accept that mathematical claims are true or false, not just true or false in application, and we have to come up with a way to understand how one mathematical claim is more or less plausible than another. It is this latter that has particularly stymied attempts at analyzing mathematical proofs as explanations.[25]

Further, for an explanation to be good, at least one of the claims doing the explaining must be no more plausible than the claim being explained. Otherwise, we would have an argument for the conclusion. Thus we would have to have that some inferences to "$1 + 2 + \cdots + n = 1/2 \, n \cdot (n+1)$" are to be judged as arguments for that claim, establishing the truth of it, and some are to be judged as explanations, telling us why the claim is true. There would not be a uniform standard by which to judge mathematical proofs.

None of these problems arise if we go back to the schematic diagram of mathematical proofs presented on p. 76 above. The mathematical inference to the mathematical claim is neither an argument nor is it an explanation; it is a pure inference, to be judged solely as to whether it is valid or not. The mathematical proof is an argument that the inference is valid. Part of the feeling that some proofs are better at showing "why a claim is true" has to do with what criteria we have for such an argument to be good.

We have seen necessary conditions for an argument to be good (p. 77). We can also say that one argument is better than another if its premises are more plausible and it is more clearly valid or strong. For some kinds of arguments, such as generalizations and analogies, more can be said about what constitutes a good or better argument. But consideration of what conditions are needed for one argument to be better than another in mathematics has been obscured by seeing it in terms of explanations.

For example, Mark Steiner in "Penrose and Platonism," says,

> Now I see no reason, except dogmatism, not to accept this story at face value: the embedding of the reals in the complex plane yields explanatory proofs of otherwise unexplained facts about the real

numbers. The explanatory power of such proofs depends on our investing the complex numbers with properties they were never perceived as having before: length and direction. p. 137

There is more than dogmatism as a motive to reject Steiner's story, for it depends on our assuming that mathematical claims have truth-values. Yet if mathematical claims have truth-values, we cannot "invest" the complex numbers with properties. Either they have those properties or they don't. In any case, mathematicians don't claim that the complex numbers have those properties: we represent them using those quantities.

To see better what Steiner might be getting at, consider the what Philip Kitcher says in "Bolzano's Ideal of Algebraic Analysis":

And as in other sciences, explanation can be extended by absorbing one theory within another. It is customary to praise scientific theories for their explanatory power when they forge connections between phenomena which were previously regarded as unrelated. Within mathematics the same is true and it has become usual to defend the "abstract" approach to mathematics by appealing to the connections which are revealed by studying familiar disciplines as instantiations of general algebraic structures. pp. 259–260

It is not clear what Kitcher means by "explanation" and "explanatory power," for we are not explaining anything in his examples. Rather, we are setting out further analogies, connecting our abstractions to show that they have instantiations we didn't previously see, and showing that some of our abstractions can be abstracted further to relate them by analogy to other abstractions. As always, good analogies help us "see" the relationships in the sides of the analogy. It's not different from marijuana compared to alcohol, or humans compared to dogs: we don't explain anything with such an analogy, but we do see common aspects and reason to similar claims based on those aspects when the differences don't matter.

We need such further analogies or instantiations of our abstractions because our abstractions have become too abstract to reason about well, or because they are so abstract we are not sure they are related to anything in experience beyond what they have been abstracted from, or because we are postulating new entities that need to be shown to have an instantiation in something less abstract, or because we have run out of ways to conceive of further progress in the area and have need of some other way to visualize the subject. All these help us understand our abstractions better. But they are not explanations.

Consider what Ernest Nagel says in *The Structure of Science* about a particular mathematical claim that needs explaining:

> Why is the sum of any number of consecutive odd integers beginning with 1 always a perfect square (for example, $1 + 3 + 5 + 7 = 16 = 4^2$) ? Here the "fact" to be explained (called the *explicandum*) will be assumed to be a claimant for the familiar though not transparently clear label of "necessary truth," in the sense that its denial is self-contradictory. A relevant answer to the question is therefore a demonstration which establishes not only the universal truth but also the necessity of the explicandum. The explanation will accomplish this if the steps of the demonstration conform to the formal requirements of logical proof and if, furthermore, the premises of the demonstration are themselves in some sense necessary. The premises will presumably be the postulates of arithmetic; and their necessary character will be assured if, for example, they can be construed as true in virtue of the meanings associated with the expressions occurring in their formulation. p. 16

The fact to be explained cannot be "the sum of any number of consecutive odd integers beginning with 1 is always a perfect square" because that is not obviously true. It doesn't need to be explained but demonstrated. The following proof of that claim does all one could hope for in making clear why the claim is "true."

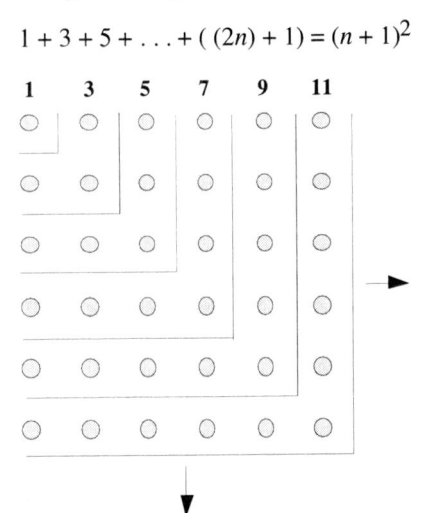

$$1 + 3 + 5 + \ldots + ((2n) + 1) = (n + 1)^2$$

92   *Reasoning in Science and Mathematics*

The diagram comes from Martin Gardner's "Mathematical Games" where he says,

> Think of the pattern as extending any desired distance to the right and down. Each reversed *L*-shaped strip contains the odd number of circles indicated at the top. It is obvious that each additional strip, that is, each new odd number in the series $1 + 3 + 5 + \ldots$, enlarges the square by one unit on a side, and that the total number of dots in each square bounded by the *n*th odd number is n^2. p. 114

The picture convinces us that the claim is true in any application of the assumptions of our theory of addition and multiplication. To say that it is thus true by virtue of the meaning of the expressions requires us to pack a very great deal into what is understood by "meaning." When a mathematical theory can be shown not to apply to some situation/experience/thing/process, such as addition to drops of water, we do not say the theory is false but that this is a bad application. Have the meanings of the expressions changed? Or are we only clarifying what we implicitly assumed were the meanings? The appropriateness of an application seems hard to assimilate to implicit meanings of expressions in a theory.

Finally, consider what Michael Resnik and D. Kushner say about proofs as explanations:

> We can account for what is probably the most basic intuition behind the idea that there must be explanatory proofs as such, namely that all proofs convince us that the theorem proved is true but only some leave us wondering why it is true. We have this intuition, we submit, because we have observed that many proofs are perfectly satisfactory as proofs but present so little information concerning the underlying structure treated by the theorem that they leave many of our why-questions unanswered. In reflecting on this, we tend to conflate these unanswered why-questions under the one form of words "why is this true?" and thus derive the mistaken idea that there is an objective distinction between explanatory and non-explanatory proofs.
>
> "Explanation, Independence, and Realism in Mathematics," p. 154

The distinction between explanatory proofs and non-explanatory ones is no different for arguments in daily life, such as newspaper editorials. Sometimes we can follow the steps of the argument very well but remain unconvinced. We are not being irrational; we just don't "see" the connections that make the argument valid or strong. Similarly, we sometimes need to "see" the connections in mathematics. It requires us not only to see the deductive connection but the relation to what we already know or to something familiar, perhaps through a higher-level abstraction that creates an analogy. Asking "Why is this true?" leads to the mistaken idea of looking for the grounds of the truth of a mathematical claim. Often what is explanatory replaces symbolic manipulation (e.g., an induction proof) with something more concrete in our experience (e.g., a picture).

The utility of this story

A good story of mathematics should lead to new and interesting work in mathematics.

By considering the path of abstracting in a mathematical subject rather than focusing only on the final abstraction, we can see where we have chosen to ignore certain aspects of our experience. Then, when the development of the subject becomes stuck, when we cannot adapt our abstractions to accommodate problems that resist solution, we can look to what we have ignored and see if it is possible to take more into account. This is the usual method of scientists; it works equally well for mathematicians.[26]

By considering how we develop and use mathematics, we are no longer left grasping at abstractions hoping that they will somehow reveal to our intellects new solutions, abstractions, or analogies through their ineffable nature. By remembering that mathematics is a human activity, we can solve more.

We can encourage this view with our teaching. By introducing the process of abstracting rather than focusing only on the final abstraction we can help students grasp new concepts and make use of them, not only as a subject to be learned but as a tool for modeling further.[27]

Grounding our stories of mathematics

The view of mathematics that I present is grounded in mathematics as a human activity. Any story of mathematics has to account for that. The assumptions about the world that I invoke, such as that there are people who do mathematics who communicate and who have intuitions, are not controversial. In that sense, I have offered a story that has minimal metaphysics.

I model what I see mathematicians doing. I do not try to account for all. Abstract things arise—or come to our attention if you are a platonist—via our abstracting. All mathematics is applied mathematics; we've just forgotten that it's applied because of its familiarity. Mathematics is abstraction, proofs, and applications, repeated over and over, limited only by our experience and imagination.

But the question can still arise: Why these abstractions?

We can ask for ultimate explanations. Perhaps we have intersubjective work in mathematics because there are platonic objects that ground our insights, or because mental constructions and subjective thoughts are somehow shared by all thinking creatures, or because . . . . These ultimate explanations are the basis of other views of mathematics. They are beyond testing, except for whether they answer the questions that a good story of mathematics should answer.

We can view various philosophies of mathematics with their accompanying metaphysics as providing us with some ultimate account of why we do mathematics as I have described. Or we can see stories of abstract objects and mental constructions as psychological props that allow us to reason better about our abstractions. No matter. Such stories are compatible with what I have said here.

As for me, I am content to do mathematics and to reflect on how we communicate when we do that. Come, let us reason together.

Appendix 1 Mathematics as an innate ability

In the text I have suggested that the counting numbers and mathematics generally arise through a process of abstraction, or at least they can be understood that way. Some psychologists agree. My discussion of counting and arithmetic on pp. 67–69 reflects my understanding of experiments done by Jean Piaget with children as described in his *The Child's Conception of Number*. There he shows how hard it is for children to learn to count and how counting and addition and subtraction are learned together.

Seymour Papert, working in that tradition, says:

> For the infant, objects do not even exist; an initial structuration is needed to organize experience into *things*. Let us stress that the baby does not *discover* the existence of objects like an explorer discovers a mountain, but rather like someone discovers music: he has heard it for years, but before then it was only noise to his ears. Having "acquired objects," the child still has a long way to go before reaching the stage of classes, seriations, inclusions and, eventually, number.
>
> "Problèmes épistémologiques et génétiques de la recurrence," as translated by Stephen Dehaene in *The Number Sense*, p. 31

That objects, "things," are part of our learned perception of the world is the thrust of my book *The Internal Structure of Predicates and Names with an Analysis of Reasoning about Process*, where I contrast that with the view of the world as process, the flux of all, suggesting that grammar is determinant of which view we take. If there is evidence that a person or creature does not have a notion of discrete but only continuous, then that is evidence that the creature cannot have a notion of counting number. And indeed there are cultures that distinguish only one, two, and many. Compare what Stephen Dehaene says in *The Number Sense*:

> Even after considerable training, a rat seems unable to press exactly four times on a lever, but it can press four, five, or six times on different trials. I believe that this is due to a fundamental inability to represent numbers 4, 5, and 6 in a discrete and individualized format, as we do. To a rat, numbers are just approximate magnitudes, variable from time to time, and as fleeting and elusive as the duration of sounds or the saturation of colors. p. 19

> First, animals can count, since they are able to increase an internal counter each time an external event occurs. Second, they do not count exactly as we do. Their representation of numbers, contrary to ours, is a fuzzy one. p. 23

To say that rats do not count exactly as we do is to say they simply don't count: a fuzzy representation of numbers is no conception of number at all—if we even had an idea of what "representing numbers" means, as opposed to registering incidents.

George Lakoff and Rafael E. Núñez dispute that arithmetic must be learned. In *Where Mathematics Comes From*, pp. 15–16, they say:

> The very idea that babies have mathematical capacities is startling. Mathematics is usually thought of as something inherently difficult that has to be taught with homework and exercises. Yet we come into life prepared to do at least some rudimentary form of arithmetic. Recent research has shown that babies have the following numerical abilities:
>
> 1. At three to four days, a baby can discriminate between collections of two and three items [reference supplied]. Under certain conditions, infants can even distinguish three items from four [reference supplied].
>
> 2. By four and a half months a baby "can tell" that one plus one is two and that two minus one is one [reference supplied].
>
> 3. A little later, infants "can tell" that two plus one is three and that three minus one is two [reference supplied].
>
> [Experiments are described in which babies stare at slides showing various numbers of objects, the length of the stare indicating to the researcher that the baby is discriminating between different configurations.]
>
> The ability to do the simplest arithmetic was established using similar habituation techniques.

Lakoff and Núñez describe it correctly in (1): babies can discriminate between collections of objects. When they say that babies can discriminate between numbers and do simple arithmetic, consider that most any mammal and many birds can distinguish between collections of two versus three objects. That does not indicate any ability to do mathematics. That does not show that such creatures or that babies have any concept of number. Abstracting the notion of a number as distinct from a number of objects is the first crucial step in mathematics, and there is no evidence that other mammals, birds, or babies can make that step.

See Peter Bryant and Terezinha Nuñes, "Children's Understanding of Mathematics," for a review of Lakoff and Núñez's view. For a survey and critique of work on the idea of mathematics as innate knowledge, see Helen De Cruz and Johan De Smedt, "The Innateness Hypothesis and Mathematical Concepts."

Appendix 2 Mathematics as used in science

Models and theories in science are presented in language, and the conventions of our languages are not meant to be part of the model; they are only the means of our communicating the ideas of the model. Part of that language is mathematics, a precise way to communicate which we have developed. Mathematics, however, is not just a language. It is a codification of inferences and methods of reasoning. When a scientist uses the calculus to deduce a claim from some theory, a great deal seems to be assumed by the mathematics over and above what is specifically codified in the theory.

The presentation of the calculus in textbooks from which scientists learn assumes that there are completed infinities, that numbers exist, and many other purely mathematical claims that on reflection would seem suspect to scientists. Indeed, the earliest uses of the calculus were severely criticized because they invoked fluxions or infinitesimals, things smaller than any real number but not zero. Yet scientists continue to use the mathematics, even though they do not assume that infinitesimals or completed infinities exist. I hope to have shown here how they may be justified in doing so.

Some, however, claim that because numbers are essential to our scientific theories, and we have good reason to believe those theories, we therefore have good reason to believe that mathematical entities exist, too. Penelope Maddy describes this *indispensability argument* in "Indispensability and Practice"[28]:

> We have good reason to believe our best scientific theories, and mathematical entities are indispensable to those theories, so we have good reason to believe in mathematical entities. Mathematics is thus on an ontological par with natural science. Furthermore, the evidence that confirms scientific theories also confirms the required mathematics, so mathematics and science are on an epistemological par as well. p. 278

But this is inference to the best explanation, reasoning backwards, which is no better in mathematics than in daily life.[29] The difference, it seems, is that in mathematics there is no other evidence we can cite for "Numbers, as abstract objects, exist." Mathematics, for the platonist, is built on faith; and the necessity of numbers for mathematics—all numbers, natural, rational, real, in their abstract plentitude—is a guide, a sign towards that faith.

The indispensability argument has an unwelcome consequence: abstract objects such as classes and numbers exist because the best science we now have requires them. But we are aware that our best science now may not be the best in the future. So these things might not exist—perhaps there are no numbers, but lots of other things, perhaps a huge different ontology.[30]

Though the assumption that numbers exist as abstract objects may be compatible with our current mathematics, we have no need of it to do mathematics and apply mathematics to science, as I hope to have shown here.[31]

Appendix 3 Comparisons to other views of mathematics

One of the virtues of the story I have told is its seeming familiarity. In the text here I've compared it to several views of mathematics that are quite different. Here I'll compare it to three well-known views of mathematics that might seem similar: structuralism, deductivism, and mathematics as an empirical science.[32]

Structuralism

Structuralism is the view that mathematics is not about objects but about relationships or, it is said, structures. Michael Resnik says:

> In mathematics, I claim, we do not have objects with an "internal" composition arranged in structures, we have only structures. The objects of mathematics, that is, the entities which our mathematical constants and quantifiers denote, are structureless points or positions in structures. As positions in structures, they have no identity or features outside a strucure.
> "Mathematics as a Science of Patterns: Ontology and Reference," p. 530

Structuralists reify structures rather than mathematical objects, and in doing so they seem to be platonists. They, too, have a problem of how mathematics can be applied. Charles Chihara says:

> I shall not attempt to explain here in any detail how geometry is applied, but a few generalities may be helpful. Let us begin with the fact that the axioms of this geometry characterize a type of structure. Since the rules of inference by which theorems are derived yield sentences that must hold in any structure that is characterized by the axioms, and since physical space itself can be represented as having a mathematical structure of the sort that is characterized by the axioms—Hilbert proves that his geometry "is identical to the ordinary 'Cartesian' geometry" [reference supplied]—it follows that the theorems proved in the geometry (when given an appropriate interpretation) must hold of the represented structure. Thus, one is justified in drawing conclusions about the lines and points constructed in physical space from the theorems of Hilbert's geometry.
> *A Structural Account of Mathematics*, p. 45

That physical space can be represented as having the structure of the axioms of some extant geometry is really the issue: What is the nature of that representation? I understand it as a process of abstraction. If we ignore enough, and that which we ignore doesn't matter to the issues involved, then we can view our experience as interpreting Euclidean geometry.

Deductivism

It is not a new idea that any substantive claim of mathematics is really an inference with the assumptions of the theory as premises and that claim as conclusion, or else is a conditional with the assumptions conjoined as antecedent and the substantive claim the consequent. Long ago Gottfried Wilhelm Leibniz said:

> As for "eternal truths", it must be understood that fundamentally they are all conditional; they say, in effect: given so and so, such and such is the case. For instance, when I say: *Any figure which has three sides will also have three angles*, I am saying nothing more than that given that there is a figure with three sides that same figure will have three angles. *New Essays on Human Understanding*, Bk IV, Ch. xi, §14

P. H. Nidditch says:

> [I]t has become more and more widely accepted during the past hundred years, with the result that it is now the orthodox doctrine, that to say of a mathematical proposition *p* that it is true is merely to say that *p* is true in some mathematical system S, and that this in turn is merely to say that *p* is a theorem in S. ... This view of the nature of mathematical truth ... was first put forward with full explicitness and clarity by the Scottish philosopher Dugald Stuart. "Whereas in all other sciences," he says, "the propositions which we attempt to establish express fact, real or supposed—in mathematics, the propositions which we demonstrate only assert a connection between certain suppositions and certain consequences. Our reasonings, therefore, in mathematics, are directed to an object essentially different from what we have in view, in any other employment of our intellectual faculties —not to ascertain *truths* with respect to actual existence, but to trace the logical filiation of consequences from our assumed hypotheses."
> *Elementary Logic of Science and Mathematics*, p. 287

Charles Sanders Peirce says,

> The most abstract of all the sciences is mathematics. That this is so, has been made manifest in our day; because all mathematicians now see clearly that mathematics is only busied about *purely hypothetical questions*. ... Mathematics does not undertake to ascertain any matter of fact whatever, but merely posits hypotheses, and traces out their consequences.
> *Collected Papers of Charles Sanders Pierce I*, p. 23 and p. 109

And Bertrand Russell says:

> Pure mathematics consists entirely of assertions to the effect that, if such and such a proposition is true of *anything*, then such and such

100   Reasoning in Science and Mathematics

another proposition is true of that thing. It is essential not to discuss whether the first proposition is really true, and not to mention what the anything is, of which it is supposed to be true. Both these points would belong to applied mathematics.
"Mathematics and the Metaphysicians," p. 75

M. Bôcher says:

The nominalism of the present day mathematician consists in treating the objects of his investigation and the relations between them as mere symbols. He then states his propositions, in effect, in the following form: If there exists any objects in the physical or mental world with relations among themselves which satisfy the conditions which I have laid down for my symbols, then such and such facts will be true concerning them.
"The Fundamental Conceptions and Methods of Mathematics," p. 122

This sounds like what I said above:

The truths of mathematics are truths about inferences.
Mathematics is about what follows from what in our abstractions.

But there are major differences.

First, deductivists now typically understand the inferences of mathematics to be either in or justified by formal logic. Here is what Hilary Putnam says:

There is another possible way of doing mathematics, however, or at any rate, of viewing it. This way, which is probably much older than the modern way, has suffered from never being explicitly described and defended. It is to take the standpoint that mathematics has *no* objects of its own at all. You can prove theorems about anything you want—rainy days, or marks on paper, or graphs, or lines or spheres—but the mathematician, on this view, makes no existence assertions at all. What he asserts is that certain things are *possible* and certain things are *impossible*—in a strong and uniquely mathematical sense of "possible" and "impossible". In short, mathematics is essentially *modal* rather than existential, on this view, which I have elsewhere termed "mathematics as modal logic".

Let me say a few things about this standpoint here.

(1) This standpoint is not intended to satisfy the nominalist. The nominalist, good man that he is, cannot accept modal notions any more than he can accept the existence of sets. We leave the nominalist to satisfy himself.
"What is Mathematical Truth?", p. 70

Putnam understands the modal nature of mathematics as justified, legitimated, or somehow essentially explicated by formal modal logic.[33] But there is nothing in our experience that can count as a possible way the world could be in which Euclidean plane geometry is true. *What counts as a possibility, and what justifies our use of inferences with claims that are neither true nor false, is an application of the theory where what we count as true or false is constrained by our agreements, explicit or not, as to what we will pay attention to in our reasoning.*

Putnam's comments about the nominalist show that his conception is based on some more ample metaphysics than mine, for there is nothing in the view I present here that prevents a nominalist from accepting it—at least in those cases where the abstraction is from sufficently clear experience. Reasoning commits us to some notion of possibility, but that notion need not be unacceptable to a nominalist.[34]

By not seeing the nature of possibility in terms of abstractions based on the same methodology as used in the sciences, deductivists are led to a kind of mystery about how mathematics can be applied. As Alan Musgrave in "Logicism Revisted" says:

> Russell sought a way to bring geometry into the sphere of logic. And he found it in what I shall call the *If-thenist manoeuvre*: the *axioms* of the various geometries do not follow from the logical axioms (how *could* they, for they are mutually inconsistent?), nor do geometrical *theorems*; but the *conditional statements linking axioms to theorems* do follow from logical axioms. Hence, geometry, *viewed as a body of conditional statements*, is derivable from logic after all. . . . Russell argued that the discovery of non-Euclidean geometries forced us to distinguish *pure geometry*, a branch of pure mathematics whose assertions are all conditional, from *applied geometry*, a branch of empirical science. pp. 109–110

> If-thenism has nothing to say about un-axiomatised or pre-axiomatised mathematics, in which creative mathematicians work. Therefore, *even if* its account of axiomatised mathematics is acceptable, as an account of mathematics as a whole it is seriously defective. p. 119

David Hilbert tried to deal with this problem:

> It is surely obvious that every theory is only a scaffolding or schema of concepts together with their necessary relations to one another, and that the basic elements can be thought of in any way one likes. If in speaking of my points I think of some system of things, e.g. the system love, law, chimney-sweep . . . and then assume all my axioms as relations between these things, then my propositions, e.g. Pythagoras' theorem, are also valid for these things. In other words: any theory can always be applied to infinitely many systems of basic elements.

> [T]he application of a theory to the world of appearances always requires a certain measure of good will and tactfulness: e.g., that we substitute the smallest possible bodies for points and the longest possible ones, e.g., light rays, for lines. We also must not be too exact in testing the propositions, for these are only theoretical propositions.
>
> David Hilbert, letter to Frege, December 29, 1899[35]

But it isn't good will and tact. The issue is the nature of abstraction and application.

A version of deductivism described by Alan Musgrave in "Logicism Revisited" seems closer to what I have presented:

> I think, for example, that the sophisticated *evolutionary* Platonism of Popper need not trouble an If-thenist. Popper tries to combine a Platonistic view of the *objectivity* of human knowledge with the Darwinian view that human knowledge is an *evolutionary product*. Thus he insists that the natural numbers are a human creation (part and parcel of the creation of descriptive languages with devices for counting things), but that once created they become *autonomous* so that objective discoveries can be made about them and their properties are not at the mercy of human whim [Popper, *Objective Knowledge*, pp. 158–161]. An If-thenist could agree with much of this. We create, first of all, languages in which to express certain empirical claims: "Two apples placed in the same bowl as two other apples give you four apples"; "Two drops of water placed together give you one bigger drop of water"; *etc*. Then we come to treat numbers and their addition in a more abstract way (so that the second statement just given does not count as an empirical refutation of "1 + 1 = 2"). This is, at bottom, to create a more or less explicit collection of 'axioms' for the natural number sequence. And then we find that, once these are granted, we must also grant other statements about numbers like "There are infinitely many prime numbers". We *discover*, in other words, that our axioms have certain unintended logical consequences. The objectivity of mathematics is guaranteed by the fact that *what follows from what* is an objective question, and we need not postulate a realm of 'abstract mathematical entities' to ensure it. p. 125

The objectivity of mathematics comes from the objectivity of the inference relation *relative to our assumed metaphysics*. Still, neither Popper nor Musgrave developed these ideas into an analysis of mathematical proof and applications of mathematics, nor did they relate this to the methodology of the sciences.

Mathematics as an empirical science
After writing the body of this text I discovered *Mathematician's Delight* by the mathematician W. W. Sawyer, who shows more clearly and thoroughly than I the development of mathematical theories by a process of abstraction. That part of my view, at least, has been commonplace among mathematicians for a long time. David Sherry in "The Role of Diagrams in Mathematical Arguments" presents an analysis of the use of diagrams in mathematics on that basis that supports the view I have presented.

John Stuart Mill is supposed by many to claim that mathematics is an induction from experience. But when he uses the words "induction" and "generalization" in discussions of mathematics he seems to mean what I mean by "abstraction." Read that way his views are very similar to mine:

> We can reason about a line as if it had no breadth; because we have a power, which is the foundation of all the control we can exercise over the operations of our minds; the power, when a perception is present to the senses, or a conception to our intellects, of *attending* to a part only of that perception or conception, instead of the whole. . . .
>
> Since, then, neither in nature, nor in the human mind, do there exist any objects exactly corresponding to the definitions of geometry, while yet that science can not be supposed to be conversant about nonentities; nothing remains but to consider geometry as conversant with such lines, angles, and figures, as really exist; and the definitions, as they are called, must be regarded as some of our first and most obvious generalizations concerning these natural objects. The correctness of these generalizations, *as* generalizations, is without a flaw: the equality of all the radii of a circle is true of all circles, so far as it is true of any one: but it is not exactly true of any circle; it is only nearly true; so nearly that no error of any importance in practice will be incurred by feigning it to be exactly true. When we have occasion to extend these inductions, or their consequences, to cases in which the error would be appreciable—to lines of perceptible breadth or thickness, parallels which deviate sensibly from equidistance, and the like—we correct our conclusions, by combining them with a fresh set of propositions relating to the aberration; just as we also take in propositions relating to the physical or chemical properties of the material, if those properties happen to introduce any modification into the result; . . .
>
> When, therefore, it is affirmed that the conclusions of geometry are necessary truths, the necessity consists only in this, that they correctly follow from the suppositions from which they are deduced.
>
> *System of Logic*, Book II, Chapter V, §1

Whether Mill's views really are similar to or even compatible with mine will have to await further study.[36]

Notes

1. (p. 68) This section summarizes "Models and theories" in this volume.

2. (p. 70) We recognize this in English with the distinction between count and non-count (mass) terms.

3. (p. 71) Benjamin Lee Whorf in "The Relation of Habitual Thought to Language" says:

> Our tongue makes no distinction between numbers counted on discrete entities and numbers that are simply "counting itself." Habitual thought then assumes that in the latter the numbers are just as much counted on "something" as in the former. This is objectification. p. 140

In a similar vein, Aristotle in *Physics*, Book IV, 223a, says:

> If there cannot be some one to count there cannot be anything that can be counted, so that evidently there cannot be number; for number is either what has been, or what can be, counted.

4. (p. 71) See *Philosophy of Mathematics and Mathematical Practice in the Seventeenth Century* by Paolo Mancosu.

5. (p. 75) See Eugene Wigner, "The Unreasonable Effectiveness of Mathematics in the Natural Sciences."

6. (p. 75) Reuben Hersh in "Inner Vision, Outer Truth" discusses this issue from a similar perspective.

7. (p. 76) See Carnielli's and my presentation in *Computability*.

8. (p. 77) Compare the presentation of Euclidean geometry in my *Classical Mathematical Logic*. Albert Einstein in *Relativity* says:

> Geometry sets out from certain conceptions such as "plane," "point," and "straight line," with which we are able to associate more or less definite ideas, and from certain simple propositions (axioms) which, in virtue of these ideas, we are inclined to accept as "true." Then, on the basis of a logical process, the justification of which we feel ourselves compelled to admit, all remaining propositions are shown to follow from those axioms, *i.e.*, they are proven. A proposition is then correct ("true") when it has been derived in the recognised manner from the axioms. The question of the "truth" of the individual geometrical propositions is thus reduced to one of the "truth" of the axioms. Now it has long been known that the last question is not only unanswerable by the methods of geometry, but that it is in itself entirely without meaning.

We cannot ask whether it is true that only one straight line goes through two points. We can only say that Euclidean geometry deals with things called "straight lines," to each of which is ascribed the property of being uniquely determined by two points situated on it. The concept "true" does not tally with the assertions of pure geometry, because by the word "true" we are eventually in the habit of designating always the correspondence with a "real" object; geometry, however, is not concerned with the relation of ideas involved in it to objects of experience, but only with the logical connection of these ideas among themselves.

It is not difficult to understand why, in spite of this, we feel constrained to call the propositions of geometry "true." Geometrical ideas correspond to more or less exact objects in nature, and these last are undoubtedly the exclusive cause of the genesis of these ideas. pp. 1–2

9. (p. 78) Though I criticize Hersh here, there is much good in his book, and reading it stimulated me to write this paper.

10. (p. 79) See the axiomatization of Euclidean plane geometry due to Leslaw Szczerba in my *Classical Mathematical Logic*. Szczerba assures me that no one has a simpler independent axiomatization.

11. (p. 81) See Brian Davies' "Whither Mathematics?" for fuller discussions of these and more examples of methods of proof that are currently in debate in the mathematical community.

12. (p. 82) See, for example, D. van Dantzig "Is $10^{10^{10}}$ a finite number?"

13. (p. 82) Or in the case of David Isles, "Remarks on the Notion of Standard Non–Isomorphic Natural Number Series" and "Questioning Articles of Faith," that arithmetical functions such as multiplication and exponentiation do not obviously lead to places on the list of numbers that we can count to by 1's.

14. (p. 82) The picture comes from *Proofs without Words* by Roger B. Nelsen, p. 69. Martin Gardner, in "Mathematical Games" from which Nelsen takes this example, says:

> The first n consecutive positive integers can be depicted by dots in triangular formation. Two such triangles fit together to form a rectangular array containing $n(n+1)$ dots. Because each triangle is half of the rectangle, we see at once that the formula for the number of dots in each triangle is $n(n+1)/2$. This simple proof goes back to the ancient Greeks. p. 114

15. (p. 82) A colleague said that the picture is convincing for $n = 6$ but not for any larger numbers. But then why not say it is good only for circles colored

and laid out in this manner? See David Sherry, "The Role of Diagrams in Mathematical Arguments" for a fuller discussion of this point.

16. (p. 84) We can easily see that the method does not depend on these specific triangles and rectangles. Indeed, in ancient times often just one example, such as using "5" and "8" here (which is the ratio as close as the printer or computer monitor can manage), was given and the reader was expected to see that the method of proof or calculation was quite general. See I. G. Bashmakova and G. S. Smirnova, "Geometry: The First Universal Language of Mathematics."

The issue of when we are justified in understanding specific parts of a diagram as variable is examined in "Main Problems of Diagrammatic Reasoning. Part I: The Generalization Problem" By Zenon Kulpa. The counterpart in reasoning with formulas is given a precise answer in classical predicate logic by the theorem on constants: if a name appears in a theorem but does not appear in the axioms, then it can be replaced by a variable that is quantified universally (Lemma 10.b of Chapter X of *Classical Mathematical Logic*). Kulpa's paper also surveys recent work on formalizing reasoning with diagrams.

17. (p. 84) See Herbert Breger "Tacit Knowledge and Mathematical Progress" for a discussion of this point and more on the nature of abstraction in mathematics. See Jeremy Avigad, Edward Dean, and John Mumma, "A Formal System for Euclid's Elements" for an analysis of Euclid's *Elements* that clarifies the role of diagrams and shows that the use of diagrams need not be inexact.

18. (p. 85) See Don Fallis' review of papers by Reuben Hersh and others.

19. (p. 85) As I show with examples in *Classical Mathematical Logic*, first-order logic is rarely appropriate because much mathematics requires second-order assumptions. Second-order logic and set-theory give no unique standard because there are many systems of those that differ too much. Moreover, it is quite common to establish a theorem in a formalized theory by semantic means rather than with a syntactic proof, as can be seen in *Classical Mathematical Logic*.

20. (p. 87) See the presentation in Carnielli's and my *Computability*.

21. (p. 87) In *Gottlob Frege: The Philosophical and Mathematical Correspondence*, pp. 39–40. Frege strongly disputes Hilbert's view in that volume (pp. 43–47).

22. (p. 88) See Hilary Putnam, "What is Mathematical Truth?", p. 73.

23. (p. 88) This issue in historical context is discussed in "Formalism" by Michael Detlefsen.

24. (p. 88) See, for example, Paolo Mancosu's *Philosophy of Mathematics and Mathematical Practice in the Seventeenth Century.*

25. (p. 89) See Paolo Mancosu, "On Mathematical Explanation."

26. (p. 93) This is how I developed the general framework for semantics in my *Propositional Logics*. Rather than viewing propositional logics as about abstract things called "propositions," I saw them as abstractions of how to reason using ordinary language. Rather than looking for the "right" logic that captures exactly the properties of abstract things, I saw that what we pay attention to in our abstracting, what aspects of ordinary language claims we deem important, determines the appropriateness of the logic we choose. As we vary the aspect we deem important, we vary the logic. That variation, I saw, can be described by devising an abstraction of our propositional logics—an abstraction of our abstractions. The general structures that arise are then worth investigating, not only for their own interest but for the relations among logics they illuminate and for the assumptions about how to reason well that they uncover. I did not deny the abstract, for how could I show that there are no such things as abstract propositions? Rather, I focused on the process of abstracting, and that gave rise to new mathematics that is grounded in experience. See "A General Framework for Semantics for Propositional Logics" in *Reasoning and Formal Logic* in this series for a summary of that work.

This is also what I do in *Classical Mathematical Logic*. I present formal logic as an abstraction of reasoning. By considering propositions as actual utterances or inscriptions rather than making the abstraction to treat different utterances as the same thing, I give a formal resolution of the liar paradox. By considering how we assign truth-values to atomic propositions rather than assuming that they come with truth-values, I develop a simple modification of classical mathematical logic that deals with names that do not refer.

27. (p. 93) In *Computability*, Walter Carnielli and I use the historical development of the theory of computable functions to recapitulate the path of abstraction in that subject. This is the "genetic" method of teaching, long favored by George Pólya, as discussed in his *Mathematical Methods in Science*.

28. (p. 97) Maddy criticizes the indispensability arguments on other grounds than I present here.

29. (p. 97) See the essay "Background" in this volume.

30. (p. 97) Maddy develops this idea in "Indispensability and Practice," pp. 285–289.

31. (p. 97) See also Solomon Feferman, "Why a Little Bit Goes a Long Way: Logical Foundations of Scientifically Applicable Mathematics."

32. (p. 98) For an overall exposition of current views of mathematics see the introduction to *The Architecture of Modern Mathematics* by J. Ferreirós and J. J. Gray .

33. (p. 100) In "Mathematics without Foundations" Putnam says that his view of mathematics as modal logic is equivalent to taking mathematics as based on set theory. In "What is Mathematical Truth?", p. 72, he says:

> The main question we must speak to is simply, *what is the point*? Given that one can either take modal notions as primitive and regard talk of mathematical existence as derived, or the other way around, what is the advantage to taking the modal notions as the basic ones?

34. (p. 101) See my "On valid inferences"

35. (p. 102) In *Gottlob Frege:The Philosophical and Mathematical Correspondence*, pp. 40–41.

36. (p. 104) John Skorupski in "Later Empiricism and Logical Positivism" discusses Mill's views and suggests a program for empiricists that is similar to what I have done here.

Bibliography

Page references are to the most recent edition or reprint cited unless noted otherwise. *Italics* in quotations are in the original unless noted otherwise.

ARISTOTLE
    1930   *Physics*
              Translated by R. P. Hardie and R. K. Gaye in *The Works of Aristotle*, ed. W. D. Ross, Clarendon Press, Oxford.

AVIGAD, Jeremy, Edward DEAN, and John MUMMA
    2009   A Formal System for Euclid's *Elements*
              The Review of Symbolic Logic, vol. 2, pp. 700–768.

BASHMAKOVA, I. G. and G. S. SMIRNOVA
    2000   Geometry: The First Universal Language of Mathematics
              In GROSHOLZ and BREGER, pp. 331–340.

BEVERIDGE, W. I. B.
    1957   *The Art of Scientific Investigation*
              Norton.

BIRNER, Jack
    2002   *The Cambridge Controversies in Capital Theory*
              Routledge.

BLAUG, Mark
    1980   Economic Methodology in One Easy Lesson
              British Review of Economic Issues, May 1980. Also in *Economic History and the History of Economics*, Blaug, New York University Press, 1986.

BÖCHER, M.
    1904–05   The Fundamental Conceptions and Methods of Mathematics
              Bulletin of the Ameican Mathematical Society, pp. 115–135.

BREGER, Herbert
    2000   Tacit Knowledge and Mathematical Progress
              In GROSHOLZ and BREGER, pp. 221–230.

BROUWER, L. E. J.
    1913   Intuitionisme en formalisme
              Wiskundig tijdschrift 9, pp. 199–201. Translated as "Intuitionism and formalism" by Arnold Dresden, *Bulletin of the American Mathematical Society*, vol. 20, pp. 81–96; reprinted in *Philosophy of Mathematics*, eds. P. Benacerraf and H. Putnam, 2nd ed., Cambridge University Press, pp. 77–89.

Bibliography

BRYANT, Peter and Terezinha NUÑES
   2002   Children's Understanding of Mathematics
            In *Blackwell Handbook of Childhood Cognitive Development*,
            ed. U. Goswami, Blackwell, pp. 412–439.

CANTOR, Georg
   1883   *Grundlagen einer allgemeinen Mannigfaltigkeitslehre*
            Teubner, Leipzig. (The translation here comes from
            Georg Cantor: His Mathematics and Philosophy of the Infinite,
            J. W. Dauben, Harvard University Press, 1979, pp. 128–129.)

CARTWRIGHT, Nancy
   1981   The Reality Of Causes in a World of Instrumental Laws
            In *The Philosophy of Science*, eds. R. Boyd, P. Gasper, and
            J.D. Trout, MIT Press, pp. 379–386.

CHIHARA, Charles S.
   2004   *A Structural Account of Mathematics*
            Oxford University Press.

CORDERO, Alberto
   2010   The Infinitely Faceted World: Intimations from the 1950s
            Analítica, Año 4, No. 4, 2010, pp. 9–26

DAVIES, Brian
   2005   Whither Mathematics?
            Notices of the American Mathematical Society, vol. 52,
            pp. 1350–1356.

DARWIN, Charles
   1962   *On the Origin of Species*
            Sixth edition. Collier.

DE CRUZ, Helen and Johan DE SMEDT
   2010   The Innateness Hypothesis and Mathematical Concepts
            Topoi, vol. 29, pp. 3–13.

DETLEFSEN, Michael
   2005   Formalism
            Oxford Handbook of Philosophy of Mathematics and Logic,
            ed. S. Shapiro, Oxford University Press, pp. 236–317.

EINSTEIN, Albert
   1915   *Relativity: The Special and General Theory*
            Fifteenth edition, 1961, Crown Publishers, Inc.

EPSTEIN, Richard L.
   1988   A General Framework for Semantics for Propositional Logics
            In *Methods and Applications of Mathematical Logic*, eds.
            W. Carnielli and L. P. de Alcantara, *Contemporary Mathematics*,
            no. 69, pp. 149–168. Revised in Epstein, *Reasoning and Formal*
            Logic, Advanced Reasoning Forum, 2013.

1990 *Propositional Logics*
Kluwer. 2nd edition, Oxford University Press, 1995. 2nd edition with corrections, Wadsworth, 2000. 3rd edition, Advanced Reasoning Forum, 2012.
1994 *Predicate Logic*
Oxford University Press. Reprinted, Wadsworth, 2000.
1998 *Critical Thinking*
with Carolyn Kernberger, Wadsworth, 3rd edition, 2005.
2006 *Classical Mathematical Logic*
Princeton University Press.
2010 *The Internal Structure of Predicates and Names with an Analysis of Reasoning about Process*
Typescript available at www.AdvancedReasoningForum.org .
2011 Valid Inferences
In *Logic without Frontiers: Festschrift for Walter Alexandre Carniellionthe occasion of his 60th Birthday* , eds. J.-Y. Béziau and M.E. Coniglio, College Publications, pp.105-129. Reprinted in Epstein, *Reasoning and Formal Logic*, Advanced Reasoning Forum, 2013.

EPSTEIN, Richard L. and Walter A. CARNIELLI
1989 *Computability: Computable Functions, Logic, and the Foundations of Mathematics*
Wadsworth & Brooks/Cole. 3rd ed., Advanced Reasoning Forum, 2008.

FALLIS, Don
1998 Review of *Hersh, 1993* et al.
The Journal of Symbolic Logic, vol. 63, pp. 1196–1200.

FERREIRÓS, José and Jeremy J. GRAY
2006 *The Architecture of Modern Mathematics*
Oxford University Press.

FEFERMAN, Solomon
1993 Why a little bit goes a long way: Logical foundations of scientifically applicable mathematics
Proceedings of the Philosophy of Science Association 1992, 2: 422–455.

FEYNMAN, Richard P.
1985 *QED: The Strange Theory of Light and Matter*
Princeton University Press.

FRANK, Philipp G.
1961 The Variety of Reasons For Acceptance of Scientific Theories
In Frank, *The Validation of Scientific Theories*, Collier Books.

Bibliography

FREGE, Gottlob
   1980   *Gottlob Frege: The Philosophical and Mathematical Correspondence*
           eds. G. Gabriel, H. Hermes, F. Kambartel, C. Thiel, and A. Veraart, University of Chicago Press.

FRIEDMAN, Milton
   1953   The Methodology of Positive Economics
           In *Essays in Positive Economics*, U. of Chicago Press, pp. 3–43.

GARDNER, Martin
   1973   Mathematical Games
           Scientific American, vol. 229 (October), pp. 114–118.

GIGERENZER, Gerd and Reinhard SELTEN
   2001   Rethinking Rationality
           In *Bounded Rationality*, Gigerenzer and Selten, eds., MIT Press, pp. 1–12.

GROSHOLZ, E. and H. BREGER, eds.
   2000   *The Growth of Mathematical Knowledge*, Kluwer.

HARDY, G. H.
   1929   Mathematical proof
           Mind, vol. 38, pp. 1–25.

HERSH, Reuben
   1993   Proving is Convincing and Explaining
           Educational Studies in Mathematics, vol. 24, pp. 389–399.
   1997   *What is Mathematics Really?*
           Oxford University Press. Reprinted 1998, Vintage.
   2006   Inner Vision, Outer Truth
           In Hersh, *18 Unconventional Essays on the Nature of Mathematics*, Springer.

HESSE, Mary B.
   1966   *Models and Analogies in Science*
           University of Notre Dame Press.

HILBERT, David
   1899   *Grundlagen der Geometrie*
           Translated as *The Foundations of Geometry* by E. J. Townshend, The Open Court Co., 1902.

ISLES, David
   1981   Remarks on the Notion of Standard Non-Isomorphic Natural Number Series
           In *Constructive Mathematics, Proceedings of the New Mexico State University Conference*, ed. F. Richman, Lecture Notes in Mathematics, no. 873, Springer-Verlag, pp. 111–134.

2004   Questioning Articles Of Faith: A Re-Creation of the History and Theology of Arithmetic
Bulletin of Advanced Reasoning and Knowledge, vol. 2, pp. 51–59.

KITCHER, Philip
1975   Bolzano's Ideal of Algebraic Analysis
Studies in the History and Philosophy of Science, vol. 6, pp. 229–269.

KITTEL, Charles, Walter D. KNIGHT, and Malvin RUDERMAN
1965   *Mechanics (Berkeley Physics Course—Volume 1)*
McGraw-Hill.

KULPA, Zenon
2009   Main Problems of Diagrammatic Reasoning. Part I: The Generalization Problem
Foundations of Science, vol. 14, pp. 75–96.

LAKOFF, George and Rafael E. NÚÑEZ
2000   *Where Mathematics Comes From*
Basic Books.

LANDSBURG, Stephen E.
1993   *The Armchair Economist*
The Free Press.

LANDSMAN, N. P.
2006   When Champions Meet: Rethinking the Bohr-Einstein debate
Studies in History and Philosophy of Modern Physics, vol. 37, pp. 212–242.

LEHRER, Jonah
2010   The Truth Wears Off
The New Yorker, December 13.

LEIBNIZ, Gottfried Wilhelm
1765   *New Essays on Human Understanding*
Translated and edited by Peter Remnant and Jonathan Bennet, Cambridge University Press, 1981.

MADDY, Penelope
1992   Indispensability and practice
Journal of Philosophy, vol. 89, pp. 275–289.

MANCOSU, Paolo
1996   *Philosophy of Mathematics and Mathematical Practice in the Seventeenth Century*
Oxford University Press.
2000   On Mathematical Explanation
In *Grosholz and Breger*, pp. 103–119.

MILL, John Stuart
    1874   *A System of Logic, Ratiocinative and Inductive, Being a Connected View of the Principles of Evidence and the Methods of Scientific Investigation*
              Eighth ed., Harper & Brothers, New York.

MUSGRAVE, Alan
    1977   Logicism Revisted
              British Journal of Philosophy of Science, vol. 28, pp. 90–127.

NAGEL, Ernest
    1961   *The Structure of Science*
              Harcourt, Brace & World. Reprinted by Hackett Publishing Company, 1979.

NEEDHAM, Paul
    2010   Micoressentialism: What is the Argument?
              Noûs, vol. 45, pp. 1–21.

NELSEN, Roger B.
    1993   *Proofs without Words*
              Mathematical Association of America.

NEWTON, Isaac
    ????   *Optica*, Part 1, Lecture 1
              Translated from the Latin in *The Optical Papers of Issac Newton*, ed. Alan E. Shapiro.

NIDDITCH, P. H.
    1960   *Elementary Logic of Science and Mathematics*
              The Free Press of Glencoe, Illinois.

ORMEROD, Paul
    1998   *Butterfly Economics*
              Pantheon.

PAPERT, Seymour
    1960   Problèmes épistémologiques et génétiques de la recurrence
              In P. Gréco, J.-B. Grize, S. Papert, and J. Piaget, eds., *Études d'Épistemologi Génétique*, vol. II *Problèmes de la construction de nombre*, Presses Universitaire de France, pp. 117–148.

PEIRCE, Charles Sanders
    1931   *Collected Papers of Charles Sanders Peirce*, vol. I
              Harvard.

PIAGET, Jean
    1952   *The Child's Conception of Number*
              Routledge & Kegan Paul. First published as *Le Genèse du Nombre chez l'Enfant*, trans. C Gattegno and F.M. Hodgson, 1941.

POINCARÉ, Henri
   1921   *The Foundations of Science*
            Translated by G. B. Halstead, The Science Press, New York.

POLYA, George
   1963   *Mathematical Methods in Science*
            Vol. IX of *Studies in Mathematics*, School Mathematics Study Group, National Science Foundation. Revised edition edited by Leon Bowden, The Mathematical Association of America, 1977.

POPPER, Karl R.
   1972   *Objective Knowledge*
            Oxford University Press.

PUTNAM, Hilary
   1967   Mathematics without Foundations
            The Journal of Philosophy, vol. LXIV. Reprinted PUTNAM, 1979A, pp. 43–59.
   1979   What is Mathematical Truth?
            In PUTNAM, 1979, pp. 60–78.
   1979A *Mathematics, Matter, and Method*
            Cambridge University Press, 2nd ed.

RESNIK, Michael
   1981   Mathematics as a Science of Patterns: Ontology and Reference
            Nous, vol. 15, pp. 529–550.
   1987   *Choices*
            University of Minnesota Press.

RESNIK, Michael and D. KUSHNER
   1987   Explanation, Independence, and Realism in Mathematics
            British Journal for the Philosophy of Science, vol. 38, pp. 141–158.

RUSSELL, Bertrand
   1925   Mathematics and the Metaphysicians
            In Russell, *Mysticism and Logic*, Longmans, Green and Co., pp. 74–96.

SAWYER, W. W.
   1946   *Mathematician's Delight*
            Pelican Books.

SCRIVEN, Michael
   1961   The Key Property of Physical Laws—Inaccuracy
            In *Current Issues in the Philosophy of Science*, eds. H. Feigl and G. Maxwell, Holt Rinehart and Winston, pp. 91–101.
   1962   Explanations, Predictions, and Laws
            In *Minnesota Studies in the Philosophy of Science*, vol. III, eds. H. Feigl and G. Maxwell, Univ. of Minnesota Press, pp. 170–230.

SHERRY, David
  2009 The Role of Diagrams in Mathematical Arguments
        Foundations of Science, vol. 14, pp. 59–74, 2009.
SKORUPSKI, John
  2005 Later empiricism and logical positivism
        In *The Oxford Handbook of Philosophy of Mathematics and Logic*, ed. S. Shapiro, Oxford University Press, pp. 51–80.
SPENCER, J., G. BODNER, and L. RICKARD
  1998 *Chemistry*
        John Wiley and Sons, Inc.
STEINER, Mark
  2000 Penrose and Platonism
        In GROSHOLZ and BREGER, pp. 133–141.
SUÁREZ, Mauricio
  2008 *Fictions in Science*
        Routledge
SUSSMAN, R. W.
  1977 Feeding behaviour of Lemur Catta and Lemur Fulvus
        In *Primate Ecology*, ed. T.H. Clutton-Brock, Academic Press.
THALER, Richard H.
  1991 *The Winner's Curse*
        The Free Press.
VAN DANTZIG, D.
  1956 Is $10^{10^{10}}$ a Finite Number?
        Dialectica, vol. 9, pp. 273–277.
WEIDNER, Robert T. and Robert L. SELLS
  1960 *Elementary Modern Physics*
        Allyn & Bacon, Inc.
WHORF, Benjamin Lee
  1941 The Relation of Habitual Thought to Language
        In *Language, Thought, and Reality: Selected writings of Benjamin Lee Whorf*, ed. John B. Carroll, MIT Press, 1956, pp. 134–159
WIGNER, Eugene
  1967 The Unreasonable Effectiveness of Mathematics in the Natural Sciences
        In Wigner, *Symmetries and Reflections*, Indiana University Press, pp. 222–237.
WILDER, R. L.
  1944 The Nature of Mathematical Proof
        The American Mathematical Monthly, vol. 51, pp. 309–323.

Index

italic page numbers indicate a quotation from that person

abstract things
  mathematical, 27–28, 71, 73, 74, 88, 93–94, 97, 98, 102, 108
  propositions, 1, 29, 48, 108
  See also platonism.
abstracting, 20, 69
  positing existence vs., 31, 35, 36, 39–40, 42, 48–49, 71, 76, 88, 96
  set theory and, 86–88
abstraction, path of. *See* path of abstraction.
acupuncture, 65
addition, 69–71, 88, 92.
  See also counting.
agreements for reasoning, 1, 14, 53, 101.
  See also *subjectivity*.
analogies, 11–12, 19–20, 47, 68–69, 90
applicability
  of a theory, 32–33, 38–39, 69
  of mathematics, 75
application of a theory, range of, 32–33, 38–39, 42, 101–102
applied mathematics, 94, 98, 100, 101
arguments, 4, 78
  associated, 15, 17
  explanations and, 14–15, 89, 92
  good, 5, 89
  guide to repairing, 8–9
  necessary conditions to be good, 7, 79
  unrepairable, 7–9
Aristotle, *105*
"as if" talk, 39–42, 50
assertion, 1
associated argument, 15
astrology, 34, 37, 40

Avigad, Jeremy, 107
axiom of choice, 79

backwards reasoning, 11, 17, 97
Bashmakova, I.G., 107
begging the question, 6
Beveridge, W.I.B., *63–64*
bias by researchers, 65
Birner, Jack, *48*
Blaug, Mark, 50
Böcher, M., *100*
Bodner, G., *24*
Breger, Herbert, 107
Brouwer, L.E.J., *86*
Bryant, Peter, 96

Cantor, Georg, *87*
Carnielli, Walter, 105, 107, 108
Cartwright, Nancy, *48*
causal claims, 43, 49, 61, 62–63
certainty, 80
Chihara, Charles, *98*
choice, axiom of, 79
claims, 1
  causal, 43, 49, 61, 62–63
  equivalent, 2
  dubious, 5
  observational, 53–55
    See also observations.
  plausible, 5–7, 79
classical economic theory, 38–43
classical propositional logic, 28–29, 34–35
coin flips, 62–63
complex numbers, 72, 79, 89–90

computers in mathematical
    proofs, 80–81, 84
conclusion, 2
confirming an explanation, 16–17, 31
confirming a theory, 31–33, 37, 97
consistency as justifies existence, 87–88
constructivism in mathematics,
    76, 82, 86, 87
contradiction, proof by, 79
convincing, 4–5
Copernicus' model of the universe,
    22–24, 36, 49–50
Cordero, Albert, 46
correspondence principle, 46–47
counting, 69–71, 76, 82–83, 88,
    95–96, 102, 105

Darwin, Charles, *17*
Davies, Brian, 106
De Cruz, Helen, 96
De Smedt, Johan, 96
Dean, Edward, 107
decline effect, 64–65
deductivism, 99–102
Dehaene, Stephen, *95*
Descartes, René, 75
Detlefsen, Michael, 107
diagrams as proofs, 82–84, 91–92, 93,
    103, 106–107
dogs, 2, 3, 5, 7–8, 10, 13, 14–15, 16,
    29, 44, 69, 75, 90
double negation, law of, 79
drawing the line fallacy, 2
dubious claims, 5

economics, theories of, 38–43
    not a science?, 55
effectiveness of mathematics,
    unreasonable? 75
Einstein, Albert, 26, 35, 49, 50, 51,
    105–106
electrical switches, 29–30

equivalent claims, 2
ESP, 60–61
ether, the 27, 31, 35, 36, 39,
    48–49, 51
ethology, 54, 56–57
Euclidean geometry, 27–28, 32,
    38, 39, 50, 72–73, 77, 79, 84,
    98, 101, 103, 105–106
evidence, 53
    inductive, 9
existence, positing vs. abstracting.
    See abstracting.
experiments
    causal reasoning, 62–65
    decline effect, 64–65
    duplicable, 54–55
    one is not enough, 64
    replicable, 54–55
explanations, 12–17
    arguments and, 14–15
    confirming, 16–17
    dependent, 15
    in mathematics, 88–93
    independent, 15
    inference to the best, 17, 97
    inferential, 12–17, 88
    mathematical proofs as, 88–93
    necessary conditions to be good, 13

fallacy, 9
    drawing the line, 2
    inference to the best
        explanation, 17
Fallis, Don, 107
Feferman, Solomon, 108
Ferreiros, J., 109
Feynman, Richard P., *51*
"follows from", 4
formal logic, 35, 85, 91, 100–101,
    107, 108
four-color problem, 80, 84
Frank, Philipp G., *49–50*

Frege, Gottlob, 107
Friedman, Milton, *33–34*, 36, 38, *39–40*, *41–42*, 51

Galileo, 23–24, 26, 54
Gardner, Martin, *92*, *106*
gas, kinetic theory of, 24–26
generalizations, 9–11
  necessary conditions to be good, 11
geometry. *See* Euclidean geometry.
Gigerenzer, Gerd, 50
Gödel, Kurt, 87
Gray, J.J., 109
group theory, 73–74, 77, 85

H_2O, 43–45
Halley, Edmund, 23
Hardy, G.H., *85–86*
Harrison, Ross, *61*
Hersh, Reuben, *78*, 105, 107
Hesse, Mary B., 47
Hilbert, David, 84, *87*, 98, *101–102*
history is not a science?, 55

if-thenism, 101–102
indispensability argument, 97
induction proofs in mathematics, 81–83
inductive evidence, 9
inference, 2
  mathematical, 78
  valid, strong, weak, 2–3
inference to the best explanation, 17, 97
inferential explanation, 12–17, 88
infinity, 50, 76, 79, 86–88, 97
integers, 71–72, 74
intuition, mathematical, 75–76, 85–86, 92
intuitionism, 86, 87
Irby, V.D. and M.S., *57–58*
irrationals, 72
irrelevant premise, 9
Isles, David, *49*, 106

Kitcher, Philip, *90*
Kittel, C., *47–48*
Kulpa, Zenon 107
Knight, W.D., *47–48*
Kushner, D., *92*

Lakoff, George, *96*
Landsburg, Stephen E., *48*
Landsman, N.P., 51
language, 1, 86
  mathematics as — of science, 26, 32, 95–96, 97
laws in science, 31
Lehrer, Jonah, *64–65*
Leibniz, Gottfried Wilhelm, *99*
logic
  classical propositional, 28–29, 34–35
  formal, 35, 85, 91, 100–101, 107, 108
  modal, 100–101, 109
  prescriptive, 29, 34–35, 39

Maddy, Penelope, *97*, 108
Mancosu, Paolo, 105, 108
Mandel, L., *59*
maximizing utility, 38, 41–42
meaning and truth, 92
Mill, John Stuart, *103*–104
modal logic, 100–101, 109
model vs. theory, 31, 48
modifying theories, 33–37
Morgan, B.L., *59*
Mumma, John, 107
Musgrave, Alan, *101*, *102*

Nagel, Ernest, *91*
necessary truths, 67–71, 77, 91, 99–100, 103.
  See also truth of mathematical claims.
Needham, Paul, 51

negative numbers, 71, 79
Nelsen, Roger B., 106
Newton, Isaac, 58–59
Newton's theory of motion, 23–24, 26–27, 34, 35, 46–47, 47–48, 69, 71
Nidditch, P.H., 99
nominalism, 100–101
number, conception of, 95–96. See also counting.
numbers exist. See abstract things, mathematical.
numbers indispensable for science, 97
Nuñes, Terezinha, 96
Nuñez, Rafael, 96

objectivity in mathematics, 85, 102. See also subjectivity.
objects/things, 95. See also abstract things.
observation, 53–55
Ormerod, Paul, 50

Palmer, Richard, 65
Papert, Seymour, 95
path of abstraction, 35, 38–43, 51
  in mathematics, 71, 72, 75, 76, 87, 93, 108
Peirce, Charles Sanders, 99
perceptual capabilities, 36–37
permutations, 73–74
Piaget, Jean, 95
platonism, 73, 94, 97, 102
plausible claim, 5–7, 79
Poincaré, Henri, 87
Pólya, George, 108
Popper, Karl, 102
population, 9–10
possibilities, 3–4, 65, 77, 80, 88, 00–101
prediction, 16, 31
premise, 2
  irrelevant, 9

prescriptive theory, 29, 30, 34–35, 39
prime numbers, proof of, 80
principle of rational discussion, 8
progress in mathematics, 84
proofs, mathematical, 76–78
  are arguments, 78–81
  as explanations, 88–93
  as intuition, 85–86
  as form of communication, 80–81
  See also formal logic.
propositions, 1
Ptolemy's model of the universe, 21–24, 36, 49–50
Putnam, Hilary, 100, 109
Pythagoras' theorem, 83–84

quantum mechanics, 26, 46, 51

random sampling, 10
range of application of a theory, 32–38, 39–40
rational agent in economic theory, 38–43, 50
realistic theories, 33–34, 38, 40, 81–82
reasoning backwards, 11, 17, 97
reasoning by abstraction, 11–12, 20
relevant information, 38, 39
relevant premise, 9
representative sample, 10–11
Resnik, Michael, 47–48, 51, 92, 98
Rickard, L., 24
ring theory, 74–75
Ruderman, M.A., 47–48
Russell, Bertrand, 99–100, 101

sample, 9
  random, 10–11
  representative, 10–11
Sawyer, W.W., 103

Scriven, Michael, *48*
Sells, Robert L., *46–47*
Selzen, Reinhard, 50
set theory, 86–88, 107, 109
sheep, 63–64
Sherry, David, 103, 107
simpler theory, 36, 49–50, 106
Skorupski, John, 109
Smirnova, G.S., 107
Spencer, J., *24*
Steiner, Mark, *89–90*
strong inference, 3
subjective?, 4
structuralism, 98
Suarez, Mauricio, 50
subjective nature of mathematics, 85–86, 94, 95–96
subjectivity in argument evaluation, 4, 7
subtraction, 71
Sussman, R.W., *56–57*

Thaler, Richard, 50
theories
 applicability, 32–33, 38–39, 69
 comparing, 33. *See also* theory, simpler.
 confirming, , 31–33, 37, 97
 model vs., 31, 48
 prescriptive, 29, 30, 34–35, 39
 range of application of, 32–38, 39–40
 realistic, 33–34, 38, 39, 81–82
 simpler, 36, 49–50, 106
 truth of?, 30–31, 48, 87

things. *See* abstract things; objects/things
transfinite ordinals, 76
transitive preferences, 38–39
truth and meaning, 92
truth of mathematical claims, 70–71, 77, 83, 87–89, 90, 91, 92–93, 99–101, 103
 consistency of mathematical claims shows, 88
truth of scientific claims, 28–31, 32, 33–37, 43–45, 48, 69
truths, necessary, 67–71, 77, 91, 99–100, 103

unreasonable effectiveness of mathematics?, 75
utility, maximizing, 38, 41–42
utterance, 1–2, 108
Utts, Jessica, *60*

valid inference, 2–3
vagueness, 2
Van Dantzig, 106
water, 43–45
 drops of, 68, 102
weak inference, 3
 subjective?, 4
Weidner, Robert T., *46–47*
Whorf, Benjamin, *105*
Wigner, Eugene, 105
Wilder, R.L., *85*

Richard L. Epstein received his B.A. summa cum laude from the University of Pennsylvania in 1969 and his Ph.D. from the University of California, Berkeley in 1973. He was a Fulbright Fellow to Brazil and a National Academy of Sciences scholar to Poland. He is the author of the textbook *Critical Thinking* as well as *Propositional Logics*, *Predicate Logic*, *Classical Mathematical Logic*, and *Degrees of Unsolvability*. He is currently the head of the Advanced Reasoning Forum in Socorro, New Mexico.

CPSIA information can be obtained at www.ICGtesting.com
Printed in the USA
LVOW05s1546300314

379542LV00022B/898/P